Spacecraft Systems Engineering

Spacecraft Systems Engineering

Editor

Cezar Dalca

scitus academics

Spacecraft Systems Engineering

Edited by **Cezar Dalca**

Printed in 2017

ISBN: 978-1-68117-174-6
Library of Congress Control Number: 2015954092

SCITUS Academics LLC,
616, Corporate Way, Suite 2, 4766,
Valley Cottage, NY 10989

www.scitusacademics.com

Notice

Reasonable efforts have been made to publish reliable data and views articulated in the chapters are those of the individual contributors, and not necessarily those of the editors or publishers. Editors or publishers are not responsible for the accuracy of the information in the published chapters or consequences of their use. The publisher believes no responsibility for any damage or grievance to the persons or property arising out of the use of any materials, instructions, methods or thoughts in the book. The editors and the publisher have attempted to trace the copyright holders of all material reproduced in this publication and apologize to copyright holders if permission has not been obtained. If any copyright holder has not been acknowledged, please write to us so we may rectify.

Preface

Systems engineering is an interdisciplinary field of engineering that emphases on how to design and manage multifaceted engineering systems over their life cycles. Systems engineering techniques are used in complex projects, such as spacecraft design, computer chip design, robotics, software integration, and bridge building. Systems engineering uses a host of tools that involves modeling and simulation, requirements analysis and scheduling to manage complexity.

Traditional systems engineering was seen as a branch of engineering in the classical sense, i.e., as applied only to physical system, such as space craft and aircraft. More recently, systems engineering has developed to a take on a broader meaning exclusively when humans were seen as an essential component of a system. Systems engineering concentrates on analyzing and eliciting customer needs and required functionality early in the development cycle, documenting requirements, then proceeding with design synthesis and system validation while considering the complete problem, the system lifecycle.

This book, Spacecraft Systems Engineering, offers the reader with inclusive coverage of the design of spacecraft and the implementation of space missions, with a wide spectrum of space applications and space science. This embraces mechanical, electrical and thermal aspects, as well as propulsion and control. This book is valuable for aerospace engineering students, and all those involved in the technical aspects of design and engineering in the space sector.

Table of Contents

Chapter 1

Effect of Spacecraft Aerodynamics and Heat Shield Characteristics on Optimal Aero Assisted Transfer

Antonio Mazzaracchio, Mario Marchetti

Astronautical, Electrical, and Energetic Engineering Department, Sapienza University of Rome, Rome, Italy

ABSTRACT

A spacecraft designed to operate in a planetary atmosphere must have an adequate heat shield to withstand the high heat fluxes and heat loads that are generated by aerodynamic heating. Very often, the mass of the thermal protection system is a significant fraction of the total mass of the vehicle. In contrast, performing maneuvers in the atmosphere, that would be very costly in terms of propellant consumption if they were performed completely outside of the atmosphere in a classic way, is a very attractive prospective technique. The advantages and disadvantages in terms of total mass spared must be determined. The mission investigated involves an aeroassisted coplanar transfer from a high to a low Earth orbit. The approach uses a combination of three propulsive impulses in space together with an aerodynamic maneuver in the atmosphere. The heat shield adopted is fully ablative, given the expected high values of the entering heat flux. The convenience of the aeroassisted maneuver and the influence of the parameters involved are evaluated in comparison to a conventional Hohmann transfer. In particular, a parametric analysis is performed by varying the following characteristics of the vehicle: aerodynamic efficiency, mass-to-surface ratio, deorbit impulse, and initial altitude of the orbit. The influence of the thermal protection system is examined by assessing the impact of the type of ablative material employed, the thermal safety factor, and the allowable temperature for the adhesive layer on the substructure. The analysis is conducted with a highly representative thermal model by coupling the dynamic and thermal analyses and using a genetic optimizer. The optimization methodology and the thermal model are completely original. The results indicate the importance of choosing low-density ablative materials, of adopting a suitable thermal safety factor, and of choosing high-performance adhesives. The optimal trajectories obtained correspond to a zero second propulsive impulse.

INTRODUCTION

One of the future objectives in space activities is to use aeroassisted orbital maneuvers, i.e., maneuvers that are carried out with the aid of the atmosphere, to satisfy the increasingly stringent constraints of the cost of space missions. Depending on the goal of the mission, aeroassisted maneuvers are performed in different ways: aerogravity assist, aerobraking, and aerocapture. To these maneuvers, one must add the re-entry maneuvers and some possible variants such as the skip re-entry technique. These maneuvers are very attractive, although they are only theoretical at this point (except for rare uses in aerobraking), because they can substantially reduce the requirements of a space mission in terms of propulsion and flight time in favor of, among other things, the possibility of housing a larger payload.

Underlying this approach is the possibility of exploiting the presence of the atmosphere of the celestial body around which one wants to operate to lower the overall energy required. In practice, one tries to execute the complete operation with the help of aerodynamics because orbital maneuvers are quite expensive in terms of propulsion, especially those outside the orbital plan. Thus, the design of a spacecraft with specific reference to atmospheric portion of flight must present an efficient aerodynamic configuration; however, the priority constraint of the overall cost of space missions must be satisfied.

From this brief introduction, it is already clear that once shown the technical feasibility of the aeroassisted maneuver, one must evaluate its convenience compared with alternative hypotheses, such as a classical purely propulsive maneuver. In fact, optimizing a spacecraft and its mission, or more specifically its trajectory for a given mission, is always a compromise between the interests of performance, security, and economics, which are almost always in mutual conflict.

Usually, minimizing the mass of the heat shield while respecting the limits of safety is a primary requirement. Indeed, any savings in terms of the mass of the thermal protection system (TPS) can be translated into an increase of the "useful" mass, namely, into an opportunity to accommodate a greater payload or extend other spacecraft subsystems. The problem becomes more complex when considering an aeroassisted maneuver and taking into account the mass of propellant still required during some of the various phases of the maneuver itself.

In this regard, a basic scheme is usually adopted in the literature to perform an aeroassisted maneuver—the use of propulsion only in space together with one or more atmospheric segments of pure aerodynamic flight—which is also considered in this work. Maximizing the benefits at the propulsive level for an aeroassisted maneuver normally requires a more intensive use of the atmospheric phase of flight. Thus, the resulting trajectory touches the denser layers of the atmosphere with longer crossing times. Therefore, a larger TPS with its relative higher mass fraction must be adopted.

A legitimate question that arises at this point is whether the resulting increase in the mass of the TPS, together with the mass of propellant required to enter the atmosphere and then to achieve the final orbit, may override the convenience of the aeroassisted maneuver compared with a classical operation, which is a purely propulsive extra-atmospheric maneuver. It seems evident that this is an optimization problem.
All of the above considerations concerning both the evaluation of the convenience of the aeroassisted approach and the search for the optimal solution and the methodological sphere itself are the basis of the research questions that motivated this work.

An original procedure was carried out jointly with the implementation of software developed by the authors to optimize the aeroassisted orbital maneuvers using a genetic algorithm (GA) with simultaneous evaluation of the optimal configuration of the associated heat shields and the coupling of the dynamic and thermal analyses. The tool was verified by comparing its results with those found in the literature. This tool, because of its level of implementation details, is suitable for the conceptual development stages of a spacecraft and its mission.

The initial intention was to evaluate the influence of various parameters—orbital, aerodynamic, and dimensional—on the feasibility and convenience of the mission. For an introductory analysis of the problem and to evaluate the importance of various factors, coplanar transfer (aerobraking) from a high Earth orbit (HEO) to a low Earth orbit (LEO), both circular, is proposed as a case study. The vehicle is a delta wing spacecraft that is protected by an ablative TPS with uniform thickness. The problem is studied in the absence of constraints on the maximum allowable entering heat flux.

Beyond the presentation of the problem in Section 1, Section 2 describes the model and the optimization procedure. The case study is presented in Section 3, and the relevant results and analyses are discussed in Section 4. Finally, Section 5 offers a summary, conclusions, and recommendations for future improvements.

MODELS AND OPTIMIZATION

The description of the models, the governing equations, and the relevant assumptions for the problem—i.e., thermal models, aerodynamic heating, atmospheric flight mechanics, and heat shield configuration—are thoroughly presented in [1] and [2]. Reference [2] can be consulted for full details of the original optimization procedure.

All of the analysis presented here was performed using the cited software developed by the authors called ATHSHO (Aeroassisted Trajectory and Heat SHield Optimization). It is important to recall that thermal analysis is performed with a one-dimensional plane model and that the adopted GA refers to a mixed one-point/two-point crossover operator together with a reproduction plan that provides a full generational replacement with elitism [3-5].

CASE STUDY

A parametric analysis was performed by varying the following characteristics of the vehicle: aerodynamic efficiency, mass-to-surface ratio, deorbit impulse, and the altitude of the initial orbit. The influence of the characteristics of the TPS was examined by assessing the type of ablative material adopted, the thermal safety factor, and the allowable temperature for the bond-line, i.e., the adhesive layer on the substructure.

Hypotheses

The main hypotheses considered are the following:

- The initial total mass of the vehicle is given.
- The attitude control is accomplished only through the angle of attack.
- The entering heat flux is unconstrained.
- The TPS is fully ablative with uniform thickness.

Vehicle

The vehicle model is a delta wing shuttle with a high L/D ratio, which is comparable in the first instance to the configuration and dimensions of the Boeing X-37A vehicle (**Figure 1**).

The dimensions, sizes, and aerodynamic characteristics of the vehicle used were taken in part from [6-8]. Other data were reasonable assumptions made by the authors. The main dimensions and characteristics of the vehicle are listed in **Table 1**, and the principal aerodynamic and propulsive parameters are listed in **Table 2**.

The values shown are those considered for the nominal reference case. Some of them may vary depending on the purpose of the study (Section 4).

Figure 1. Boeing X-37A (source NASA.gov).

Table 1. Vehicle dimensions and characteristics

Vehicle length	l_{ve}	9.38 m
Vehicle body radius	r_b	1.00 m
Vehicle wing span	ws_{ve}	4.50 m
Vehicle wing cord	wc_{ve}	3.50 m
Vehicle reference surface	S	11.69 m^2
Vehicle TPS total surface	$S_{TPS,ve}$	42.65 m^2
Bond-line limit temperature	$T_{BL,lim}$	450 K
Thermal safety factor	TSF	1

Table 2. Vehicle's aerodynamic and propulsive characteristics

Zero-lift drag coefficient	C_{D0}	0.032
Induced drag factor	K_D	1.4
Lift coefficient derivative	$C_{L,\alpha}$	0.5699
Maximum lift coefficient	$C_{L,\max}$	0.4
Propellant specific impulse	I_{sp}	310 s

Mission and Maneuver

Table 3 lists the values of the altitudes for the initial Geostationary Earth Orbit (GEO) and final LEO for the aerobraking maneuver as well as the conventional height assumed for the atmosphere.

Even in this case, the indicated values are those used for the reference case. In particular, in the parametric analysis performed, the value of the initial HEO altitude was varied. The physical properties of the atmosphere were derived from the model 1976 US Standard Atmosphere [9].

Figure 2 shows a classic schematic for a HEO-LEO aerobraking maneuver. The strategy involves the combined use of aerodynamic maneuvering in the atmosphere and some extra-atmospheric propulsion phases. More precisely, one assumes that the propulsive phases are concentrated in three impulses in space and that the portion of atmospheric flight is performed without the use of the propulsion system. The first propulsive impulse (deorbit) at the altitude H_A of the initial HEO varies the vehicle's speed by ΔV_1 to enter the atmosphere along an elliptic orbit segment. The second impulse (boost) is applied upon the exit from the atmosphere to achieve the final LEO by ascending once more along an elliptic orbit segment. The boost is expressed by a speed variation of the vehicle, ΔV_2.

Table 3. Maneuver characteristics

Initial HEO (GEO) altitude	H_A	35,786 km
Final LEO altitude	H_B	480 km
Atmosphere's upper limit	H_{atm}	129.6 km

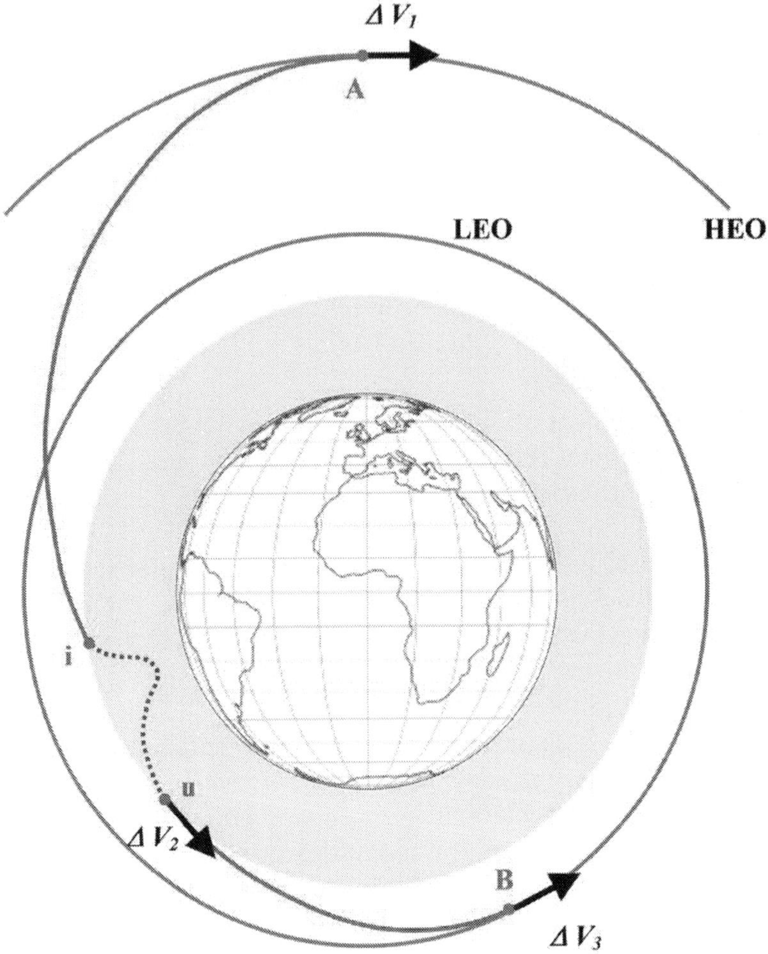

Figure 2. Schematic of a HEO-LEO aerobraking maneuver.

The third and final impulse (circularizing) varies the speed by ΔV_3 to circularize the vehicle's path within the altitude H_B of the final LEO.
It is useful at this point to describe the various phases of the maneuver in more detail as follows. Initially, the vehicle is moving on a circular orbit of radius R_A with a speed V_A around the Earth, which has radius $R_\oplus$. The expression of the circular speed is the following:

$$V_A = \sqrt{\mu / R_A} \tag{1}$$

where

$$R_A = R_\oplus + H_A. \tag{2}$$

The deorbit is accomplished by applying the first impulse ΔV_1 in the opposite direction of the spacecraft's speed. This impulse puts the vehicle along an elliptic orbit with the perigee inside the dense layers of the atmosphere.

The atmospheric region below the altitude H_{atm} in which the aerodynamic effects are considered to be conventionally present is denoted the sensible atmosphere. One can determine the speed V_i and the flight path angle γ_i of the vehicle's trajectory at the atmospheric entry point ($H = H_{atm}$) through the following relationships, which are obtained according to energetic considerations:

$$V_i = \sqrt{2\mu\left(\frac{1}{R_{atm}} + \frac{(V_A - \Delta V_1)^2}{2} - \frac{1}{R_A}\right)} \tag{3}$$

$$\gamma_i = \cos^{-1}\left[\frac{R_A}{R_{atm} V_i}(V_A - \Delta V_1)\right] \tag{4}$$

where

$$R_{atm} = R_\oplus + H_{atm}. \tag{5}$$

Clearly, it is necessary that the applied ΔV_1 be greater than the minimum $\Delta V_{1,\min}$ for which there would be only a tangential trajectory to the edge of the sensible atmosphere. This $\Delta V_{1,\min}$ is presented in **Figure 3** and given by the following expression:

$$\Delta V_{1,\min} = \sqrt{\frac{\mu}{R_A}} - \sqrt{2\mu \frac{\frac{1}{R_{atm}} - \frac{1}{R_A}}{\left(\frac{R_A}{R_{atm}}\right)^2 - 1}}. \tag{6}$$

During the atmospheric portion of flight, the vehicle performs the required maneuver, which is optimally controlled by modulations of the angle of attack α (leaving the bank angle σ fixed) subject to the heating constraints, if any. During this phase, the vehicle's speed decreases because of aerodynamic drag, and because of this loss of energy, a new impulse is necessary to achieve the final altitude.
At the end of atmospheric flight, the vehicle is situated at an altitude H_{atm} again, is driven at a speed V_u, and has a flight path angle equal to γ_u. At this moment, a boost impulse is applied to enter an ascending elliptic orbit with the apogee equal to the radius of the final circular orbit.

The required ΔV_2, as a function of V_u and γ_u, can be found from the following expression:

$$\Delta V_2 = \sqrt{2\mu \frac{\frac{1}{R_B} - \frac{1}{R_{atm}}}{1 - \left(\frac{R_B}{R_{atm}}\right)^2 \cos^2 \gamma_u}} - V_u \tag{7}$$

where

$$R_B = R_{\oplus} + H_B . \tag{8}$$

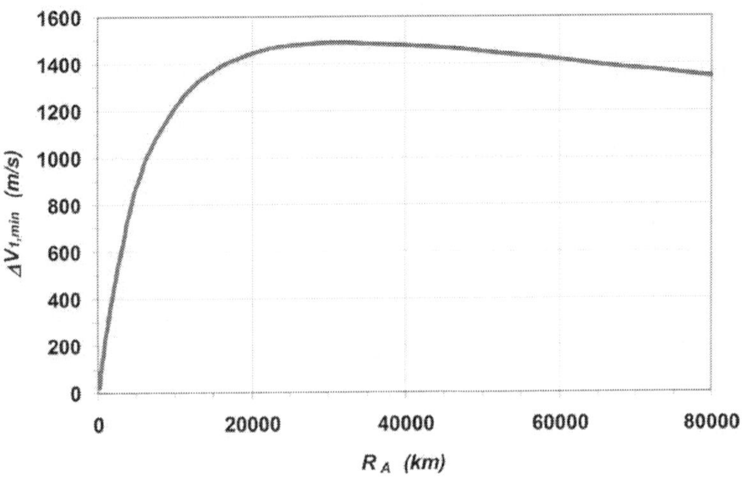

Figure 3. $\Delta V_{I,min}$ vs initial HEO radius.

Once the final altitude is reached, the third impulse is applied to circularize the final orbit. The expression for the third ΔV is the following:

$$\Delta V_3 = \sqrt{\frac{\mu}{R_B}} - \sqrt{2\mu \frac{\frac{1}{R_B} - \frac{1}{R_{atm}}}{1-\left(\frac{R_B}{R_{atm}}\right)^2 \cos^2 \gamma_u}} \cdot \frac{R_B}{R_{atm}} \cos \gamma_u . \tag{9}$$

Fitness Function, Objective Function, and Constraints

The total initial mass of the vehicle is the sum of the propellant mass, the TPS mass, and the structural and payload masses. The goal of the current optimization problem is to perform the assigned orbital transfer while minimizing the sum of the mass of the propellant and the mass of the TPS needed. The objective function, which must be maximized, is then given by the final mass of the vehicle, which can be defined as the performance index of the problem. There are some events that cause variation in the mass of the vehicle. The first is the consumption of fuel due to the deorbit impulse; thus, one can calculate the mass of the vehicle entering the atmosphere. The latter can be derived directly using the Tsiolkovsky equation for the impulse in question:

$$m_{ve,i} = m_{ve,ini}\, e^{-\frac{|\Delta V_1|}{g_0 I_{sp}}} . \tag{10}$$

This mass is further reduced by $m_{TPS,los}$ during the passage through the atmosphere because of TPS ablation. The mass loss due to ablation comes from both the surface recession and the material density change due to pyrolysis. Thus, the mass of the vehicle at the atmosphere exit is the following:

$$m_{ve,u} = m_{ve,i} - m_{TPS,los} . \tag{11}$$

At this point, the boost and circularization impulses are applied in sequence, and the final vehicle mass is obtained by two successive applications of the Tsiolkovsky equation:

$$m_{ve,fin} = m_{ve,u}\, e^{-\frac{|\Delta V_2|+|\Delta V_3|}{g_0 I_{sp}}} . \tag{12}$$

Compliance with the eventual constraint on the entering heat flux is ensured through a reward factor that is added to the objective function ($m_{ve,\,fin}/m_{ve,\,ini}$) by means of an appropriate multiplicative weight w_{HF}, which is chosen specifically to achieve rapid convergence. Likewise, the objective function is multiplied by its weight w_m. Thus, the final expression for the fitness function for the genetic optimizer is as follows:

$$ff = w_m \frac{m_{ve,\,fin}}{m_{ve,\,ini}} + w_{HF} R_{f,\,HF}. \tag{13}$$

In this case study, the heat flux is considered unconstrained, and assuming $w_m = 1$, the previous expression reduces to the following:

$$ff = \frac{m_{ve,\,fin}}{m_{ve,\,ini}}. \tag{14}$$

That is, the fitness function coincides with the objective function.

Assessment of the Convenience

The convenience of the aeroassisted maneuver can be assessed with respect to an equivalent Hohmann transfer. Such a transfer is completely "all propulsive" outside the atmosphere and based on two impulsive velocity changes at two points with radii R_A and R_B, respectively:

$$\Delta V_A = \sqrt{\frac{\mu}{R_A}}\left(\sqrt{\frac{2R_B}{R_A + R_B}} - 1\right) \tag{15}$$

$$\Delta V_B = \sqrt{\frac{\mu}{R_B}}\left(1 - \sqrt{\frac{2R_A}{R_A + R_B}}\right). \tag{16}$$

Thus, the total "all propulsive" impulse is:

$$\Delta V_{ap} = |\Delta V_A| + |\Delta V_B|. \tag{17}$$

Consequently, the expression for the final mass of the vehicle in the case of the Hohmann transfer is as follows:

$$m_{ve,fin,ap} = m_{ve,ini}\, e^{-\frac{|\Delta V_{ap}|}{g_0 I_{sp}}}. \tag{18}$$

RESULTS AND ANALYSIS

The case study for the HEO-LEO coplanar transfer through aerobraking, which was described in the previous Section 3, has been addressed by parameterising the mass-tosurface ratio M/S of the vehicle, the ΔV_1 corresponding to the deorbit impulse applied, the aerodynamic efficiency ε (with $C_{L,\max}$ fixed), the ablative material used, the maximum allowable temperature for the bondline $T_{BL,\lim}$, and the thermal safety factor TSF. In addition, the above parametric analysis was performed for different values of the initial HEO altitude, which are expressed as fractions of the GEO.

More specifically, the following values were adopted:

- HEO Altitude: 35,786 km (=GEO); 53,679 km (HEOH = 1.5 GEO); 17,893 km (HEOL = 0.5 GEO). The altitude for the final LEO is the same for all cases and equal to 480 km.
- M/S: 200 kg/m^2, 300 kg/m^2 and 450 kg/m^2, corresponding, respectively, to low, medium and high load values. Incidentally, 300 kg/m^2 is the value found most frequently in the literature. These load values correspond to the initial mass of the vehicle: 2338.0 kg, 3507.0 kg, and 5260.5 kg.
- ΔV_1 range: the range of the parametric first variation of V (in steps of 10 m/s) between the minimum value (rounded to the nearest ten higher) given by Equation (6) and the maximum value that allows a solution to complete the aeroassisted maneuver. The minimum values of the deorbit impulse for the three HEO altitudes analyzed are (**Figure 3**) 1490 m/s for GEO, 1440 m/s for HEOH, and 1420 m/s for HEOL.

- ε: 1.5 and 3 correspond, respectively, to a medium and a high aerodynamic efficiency. These values were obtained by varying C_{D0} and K_D.
- TFS: 1 and 2 correspond, respectively, to a minimum nominal thickness of the TPS and a "safer" double thickness.
- $T_{BL,\lim}$: 176.85˚C (450 K) and 400˚C (673.15 K) were both acquired from data in the literature; the first is used for standard adhesives, and the second is used for high-performance adhesives.
- Existing adhesives with enhanced characteristics allow the threshold temperature to be increased to 400˚C, which results in greater heat shield thickness savings [10].
- Ablative material: PICA-15 (hereinafter referred to as PICA), AVCOAT 5026-HCG (hereinafter referred to as AVCOAT), and FM 5055 CP in its Reduced Density version (hereinafter referred to as RDCP).

Among all the possible combinations of these parameters with their associated values, 12 cases were analyzed for each value of the M/S ratio for a total of 36 different scenarios. The combinations analyzed are presented in Tables 4-6, where the index (a, b, and c) marks the adopted value for the load. Moreover, because the range of variation of ΔV_1 is analyzed by steps of 10 m/s, 390 different cases form the database of results for the analysis of the case study in question.
As a reference, the nominal case is chosen for the GEOHEO transfer that provides a TPS made of PICA, with $T_{BL,\lim}$ = 450 K, TSF = 1, and ε = 1. The twelve combinations chosen and the corresponding purpose of

Table 4. HEO-LEO transfer, parameters and values (cases 1, 2, 3, and 4)

		Case 1a	Case 1b	Case 1c	Case 2a	Case 2b	Case 2c	Case 3a	Case 3b	Case 3c	Case 4a	Case 4b	Case 4c
HEO-LEO													
HEO = GEO	35,786 km - 480 km	■	■	■	■	■	■	■	■	■	■	■	■
M/S													
200 kg/m²	$M_{ve,int}$ = 2338.0 kg	■			■			■			■		
300 kg/m²	$M_{ve,int}$ = 3507.0 kg		■			■			■			■	
450 kg/m²	$M_{ve,int}$ = 5260.5 kg			■			■			■			■
Aerodynamic efficiency													
Medium 1.5	C_{D0} = 0.1; K = 1.111	■	■	■	■	■	■				■	■	■
High 3.0	C_{D0} = 0.017; K = 1.76							■	■	■			
ΔV_1													
	1490 - 1590 m/s	■											
	1490 - 1560 m/s		■										
	1490 - 1520 m/s			■									
	1490 - 1590 m/s				■								
	1490 - 1550 m/s					■							
	1490 - 1520 m/s						■						
	1490 - 1710 m/s							■					
	1490 - 1600 m/s								■				
	1490 - 1560 m/s									■			
	1500 - 1600 m/s										■		
	1490 - 1560 m/s											■	
	1490 - 1520 m/s												■
Ablative material													
	PICA-15	■	■	■	■	■	■	■	■	■	■	■	■
T-bond line													
	450.00 K	■	■	■				■	■	■	■	■	■
	673.15 K				■	■	■						
Thermal safety factor													
	1	■	■	■	■	■	■	■	■	■			
	2										■	■	■

Table 5. HEO-LEO transfer, parameters and values (cases 5, 6, 7, and 8).

		Case 5a	Case 5b	Case 5c	Case 6a	Case 6b	Case 6c	Case 7a	Case 7b	Case 7c	Case 8a	Case 8b	Case 8c
HEO-LEO													
HEO = GEO	35,786 km - 480 km	■	■	■	■	■	■						
HEOH = 1.5 GEO	53,679 km - 480 km							■	■	■	■	■	■
M/S													
200 kg/m²	$M_{ve,int}$ = 2338.0 kg	■			■			■			■		
300 kg/m²	$M_{ve,int}$ = 3507.0 kg		■			■			■			■	
450 kg/m²	$M_{ve,int}$ = 5260.5 kg			■			■			■			■
Aerodynamic efficiency													
Medium 1.5	C_{D0} = 0.1; K = 1.111	■	■	■	■	■	■	■	■	■	■	■	■
ΔV_1													
	No solution	■											
	No solution		■										
	1500 - 1640 m/s			■									
	1490 - 1570 m/s				■								
	1490 - 1520 m/s					■							
	1490 - 1500 m/s						■						
	1440 - 1490 m/s							■					
	1440 - 1470 m/s								■				
	1440 - 1460 m/s									■			
	1440 - 1480 m/s										■		
	1440 - 1460 m/s											■	
	1440 - 1450 m/s												■
Ablative material													
	PICA-15							■	■	■			
	FM 5055 RDCP	■	■	■									
	AVCOAT 5026-H CG				■	■	■				■	■	■
T-bond line													
	450.00 K	■	■	■	■	■	■	■	■	■	■	■	■
Thermal safety factor													
	1	■	■	■	■	■	■	■	■	■	■	■	■

Table 6. HEO-LEO transfer, parameters and values (cases 9, 10, 11, and 12).

		Case 9a	Case 9b	Case 9c	Case 10a	Case 10b	Case 10c	Case 11a	Case 11b	Case 11c	Case 12a	Case 12b	Case 12c
HEO-LEO													
HEOH = 1.5 GEO	53,679 km - 480 km	■	■	■	■	■	■						
HEOL = 0.5 GEO	17,893 km - 480 km							■	■	■	■	■	■
M/S													
200 kg/m^2	$M_{ve,tot}$ = 2338.0 kg	■			■			■			■		
300 kg/m^2	$M_{ve,tot}$ = 3507.0 kg		■			■			■			■	
450 kg/m^2	$M_{ve,tot}$ = 5260.5 kg			■			■			■			■
Aerodynamic efficiency													
Medium 1.5	C_{D0} = 0.1; K = 1.111							■	■	■			
High 3.0	C_{D0} = 0.017; K = 1.76	■	■	■	■	■	■				■	■	■
ΔV_1													
	1440 - 1530 m/s	■											
	1440 - 1490 m/s		■										
	1440 - 1460 m/s			■									
	1440 - 1520 m/s				■								
	1440 - 1470 m/s					■							
	1440 - 1450 m/s						■						
	1420 - 1880 m/s							■					
	1420 - 1640 m/s								■				
	1420 - 1550 m/s									■			
	1420 - 1880 m/s										■		
	1420 - 1810 m/s											■	
	1420 - 1620 m/s												■
Ablative material													
		■	■	■				■	■	■	■	■	■
	AVCOAT 5026-HCG				■	■	■						
T-bond line													
	450.00 K	■	■	■	■	■	■	■	■	■	■	■	■
Thermal safety factor													
	1	■	■	■	■	■	■	■	■	■	■	■	■

The evaluation are summarized in **Table 7**. Figures 4 to 15 report the trends of the mass gain compared with the "all propulsive" case with varied applied ΔV_1 and parameterized as a function of the three chosen values of M/S, which were analyzed for each scenario.

The graph on the left of each figure shows the cited mass gain in absolute terms, whereas the graph on the right presents its performance in terms of the percentage of the initial vehicle's mass.

The mass gain is therefore a measure of the convenience of the aeroassisted maneuver compared with the corresponding Hohmann transfer. When reading these graphics, the definitions of the "range of feasibility" and the "range of convenience" are introduced.

The first one is the interval of ΔV_1 in which the aeroassisted maneuver is feasible, whereas the second ΔV_1 interval is the range in which the aeroassisted operation is more convenient than the "all propulsive" one. As mentioned above, having made the run for all values of ΔV_1 leading to a solution, the range of feasibility is directly readable as the x-axis interval of the definition of the curves themselves.

The range of convenience is by definition the interval in which the curve takes positive values. The range of convenience is always within the range of feasibility. **Table 8** shows, for the value M/S = 200 kg/m^2 (case "a" of Tables 4-6), the values of the three propulsive impulses applied in each scenario, which in each case give the highest convenience for the maneuver (maxima of the "blue" curves in the graphs of Figures 4 to 15).
It is evident that the utmost convenience for the aeroassisted maneuver corresponds to a value of ΔV_2 that is very close to zero, i.e., in the absence of the boost impulse (case 5a has no feasible solutions for the considered value for the load).

This outcome is valid in general for the other values of M/S. Thus, the trajectory optimization process tends to reduce the vehicle's speed while exiting the atmosphere to the "right" value to reach the final LEO with minimal energy expenditure. Saving this fuel for the second impulse is the basis for the achievement of the highest convenience.

Comparing Figures 4 to 15, one can draw some interesting conclusions:

- The increase of M/S in all cases involves a decrease in the range of feasibility; in particular, the maximum possible ΔV_1 diminishes (the right bound of the range).
- The increase of $T_{BL,\lim}$ (**Figure 5**) from the standard value to the high-performance value due to the reduced thickness of the TPS improves the maximum achievable convenience (by approximately 2%).

Table 7. HEO-LEO transfer, scenarios analyzed.

	Combination	Purpose
1	GEO; PICA; ε medium; $T_{BL,\lim}$ standard; TSF = 1	Reference nominal case
2	GEO; PICA; ε medium; $T_{BL,\lim}$ high performance; TSF = 1	$T_{BL,\lim}$ influence
3	GEO; PICA; ε high; $T_{BL,\lim}$ standard; TSF = 1	ε influence
4	GEO; PICA; ε medium; $T_{BL,\lim}$ standard; TSF = 2	TSF influence
5	GEO; RDCP; ε medium; $T_{BL,\lim}$ standard; TSF = 1	Material influence
6	GEO; AVCOAT; ε medium; $T_{BL,\lim}$ standard; TSF = 1	Material influence
7	HEOH; PICA; ε medium; $T_{BL,\lim}$ standard; TSF = 1	HEO altitude influence; reference case for HEOH
8	HEOH; AVCOAT; ε medium; $T_{BL,\lim}$ standard; TSF = 1	Material influence w.r.t. HEOH
9	HEOH; PICA; ε high; $T_{BL,\lim}$ standard; TSF = 1	ε influence w.r.t. HEOH
10	HEOH; AVCOAT; ε high; $T_{BL,\lim}$ standard; TSF = 1	Material + ε influence w.r.t. HEOH
11	HEOL; PICA; ε medium; $T_{BL,\lim}$ standard; TSF = 1	HEO altitude influence; reference case for HEOL
12	HEOL; PICA; ε high; $T_{BL,\lim}$ standard; TSF = 1	ε influence w.r.t. HEOL

Legend: in red: GEO nominal case and changes w.r.t. it; in blue: HEOH nominal case and changes w.r.t. it; in green: HEOL nominal case and changes w.r.t. it.

Table 8. Propulsive impulses in the more convenient maneuvers for "a" cases

Case	ΔV_1(m/s)	ΔV_2 (m/s)	ΔV_3 (m/s)
1a	1530.00	1.38	334.75
2a	1530.00	-1.83	332.91
3a	1500.00	-2.21	375.26
4a	1520.00	0.69	526.53
5a	-	-	-
6a	1520.00	-0.95	502.54
7a	1480.00	2.86	366.44
8a	1470.00	-4.00	568.53
9a	1450.00	-3.51	423.53
10a	1450.00	0.67	421.13
11a	1460.00	-3.56	410.11
12a	1440.00	-2.73	267.69

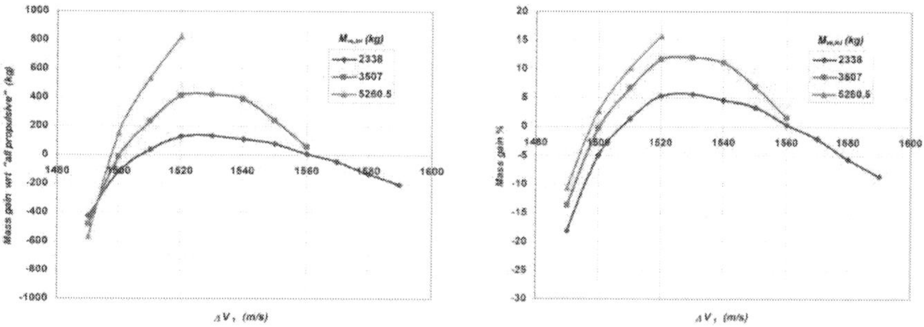

Figure 4. HEO-LEO scenario 1: reference nominal case.

- The increase in the aerodynamic efficiency of the vehicle (**Figure 6**) involves a significant extension of the range of feasibility, but a more moderate increase in the range of convenience. Cases with higher M/S benefit most. Even here, the maximum achievable convenience increases (approximately 2% more).
- The doubling of the thermal safety factor (**Figure 7**), which increases the mass of the TPS, represents a significant performance penalty for the aeroassisted maneuver, which remains affordable for a short interval of ΔV_1 only for the case with highest M/S.
- The use of other ablative materials (Figures 8 and 9), compared with the reference material PICA, appears to be inappropriate for a GEO initial altitude. Using the RDCP, the sole case with the highest M/S is feasible, but still not convenient. Conversely, using the AVCOAT increases the feasibility compared with RDCP but limits convenience compared with PICA. Thus, the relatively low density of the PICA is a discriminating element in the selection of the ablative material. Actually, these results are conservative because of the assumption that the vehicle is covered with a TPS with a uniform initial thickness.

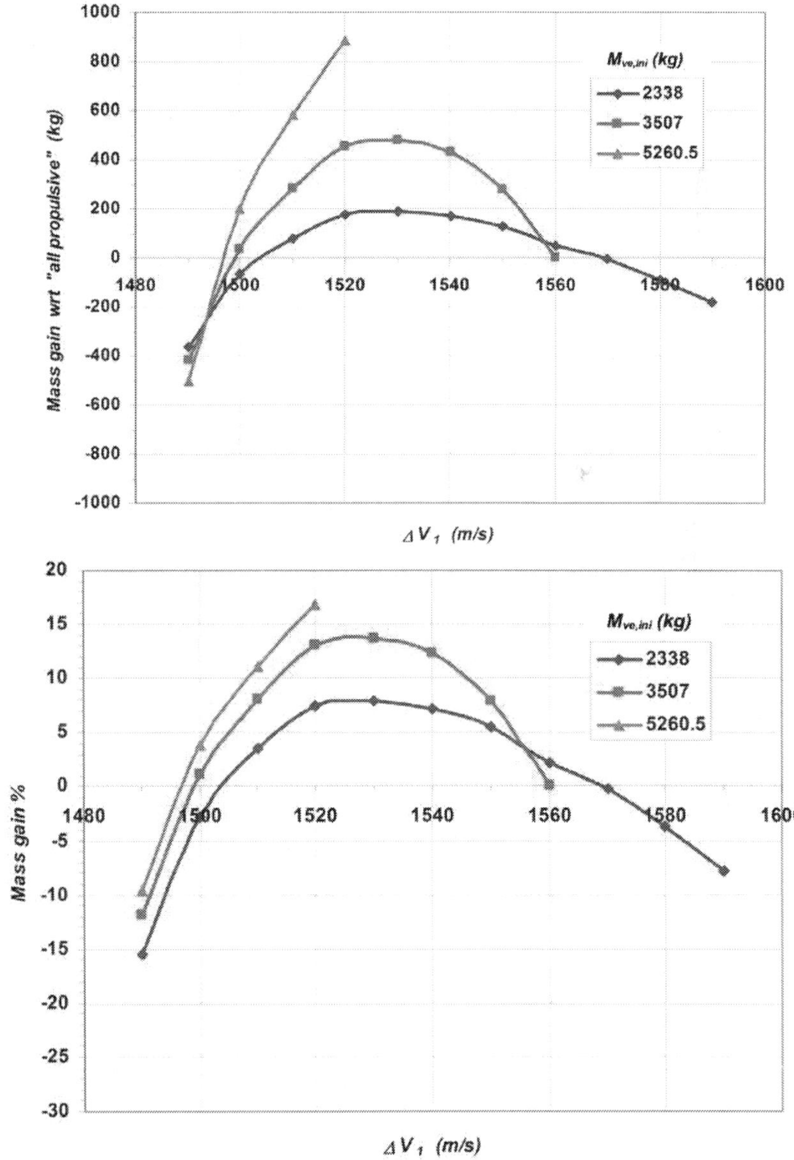

Figure 5. HEO-LEO scenario 2: $T_{BL,lim}$ influence.

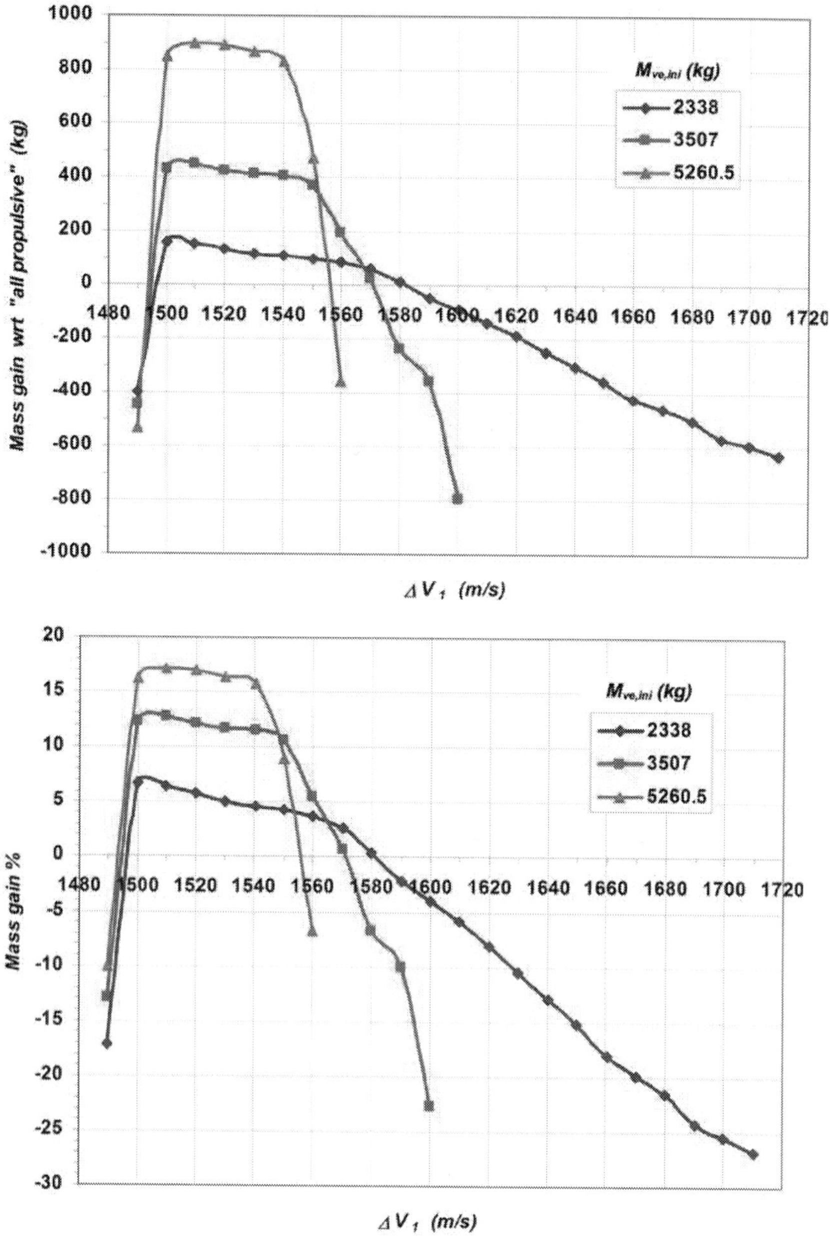

Figure 6. HEO-LEO scenario 3: e influence.

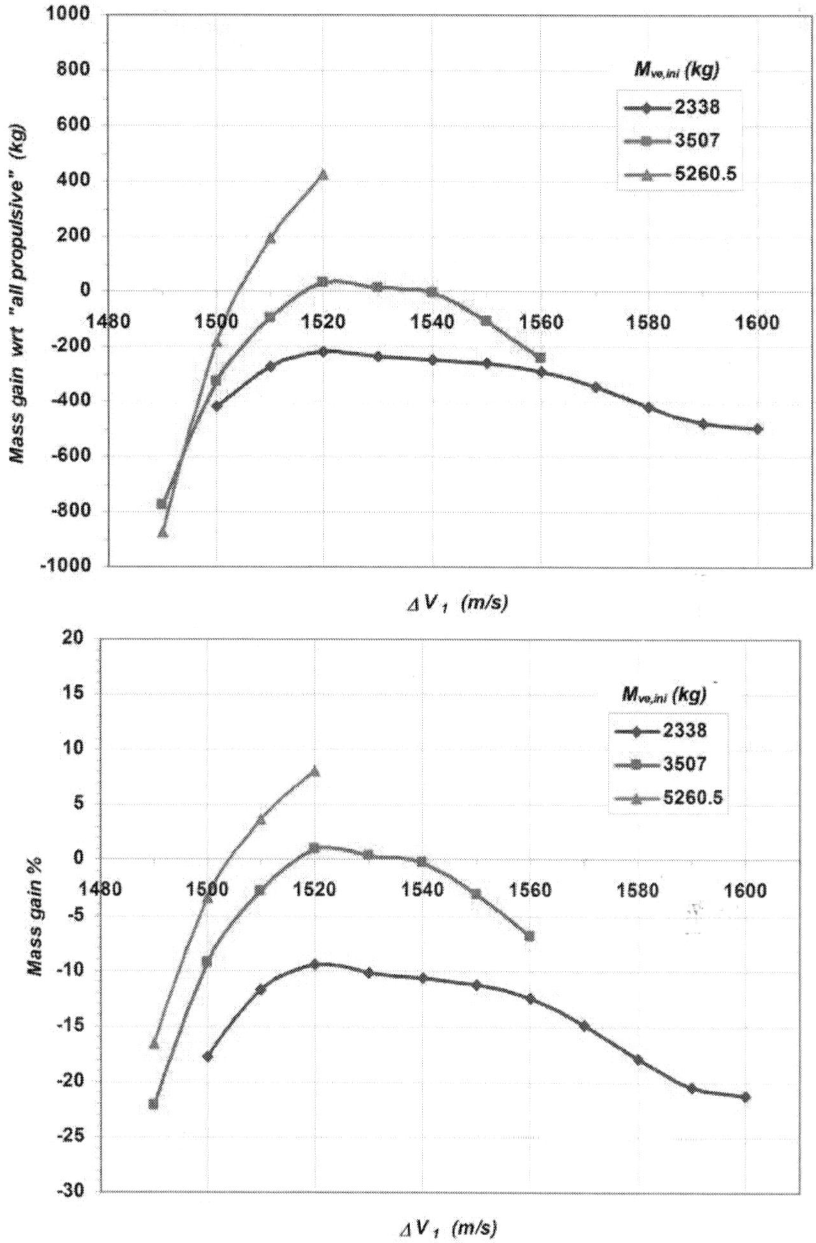

Figure 7. HEO-LEO scenario 4: TSF influence.

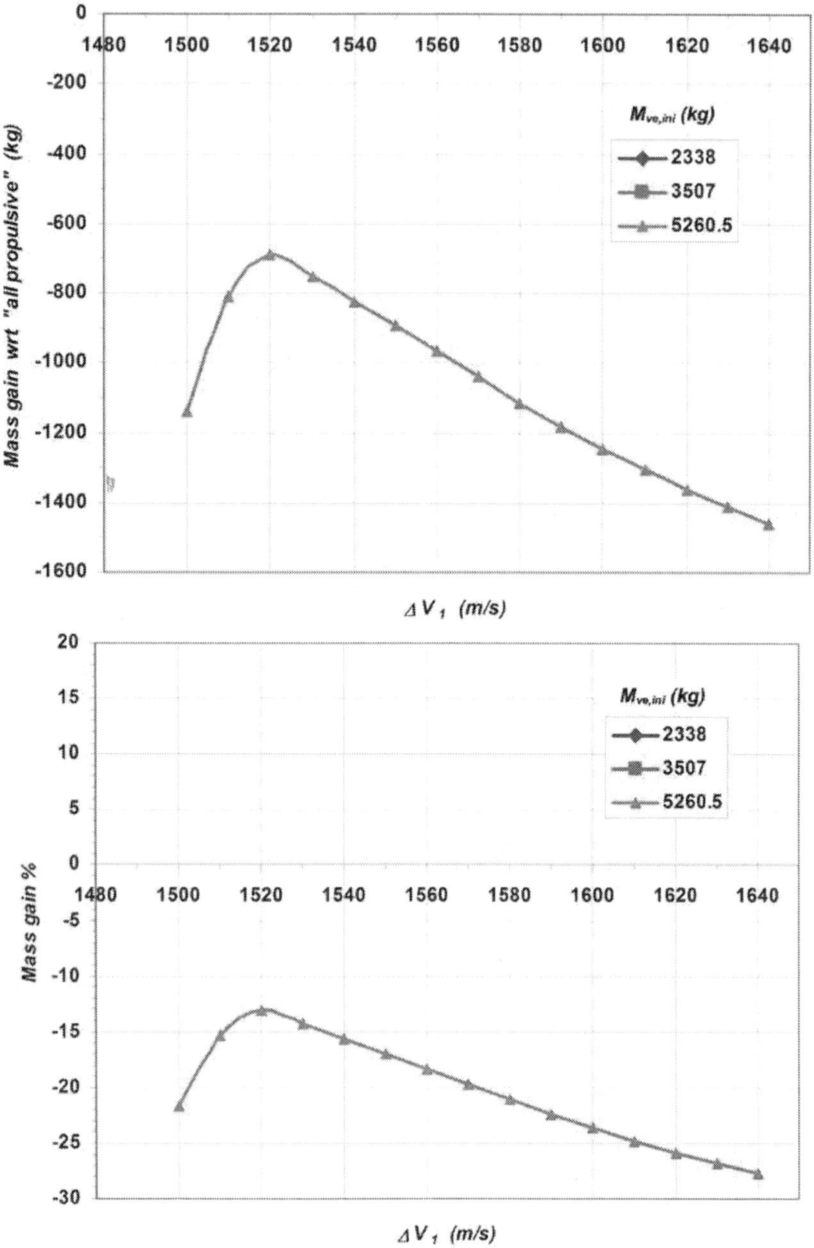

Figure 8. HEO-LEO scenario 5: material (RDCP) influence.

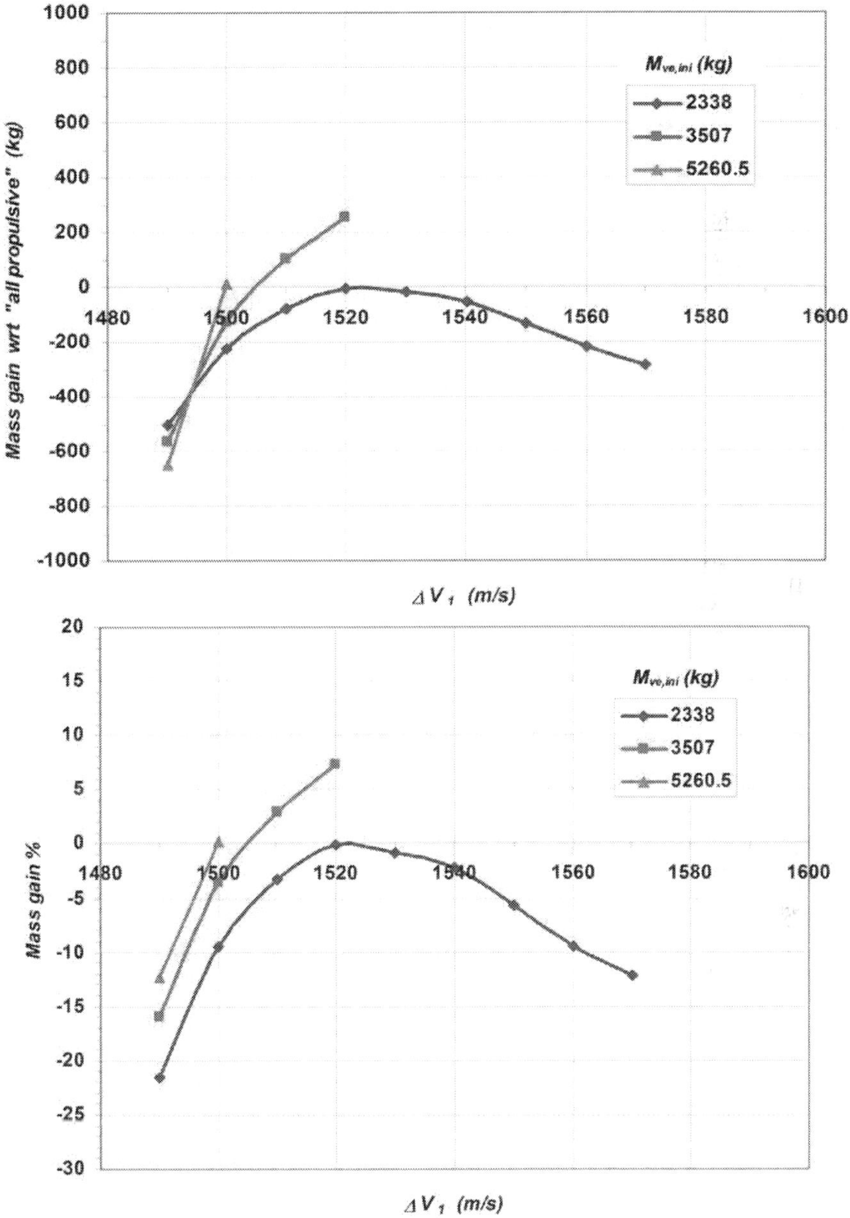

Figure 9. HEO-LEO scenario 6: material (AVCOAT) influence.

In **Figure 8**, all runs with load values of 200 kg/m^2 and 300 kg/m^2 give results, but the curves are not represented because the final "useful" masses (i.e., all the masses that are not propellant and TPS) are negative.

Thus, the vehicle is able to get out of the atmosphere, but without enough fuel to complete the mission.

- Figures 10 (HEO high altitude) and 14 (HEO low altitude) show that the aeroassisted maneuver is much more convenient if the HEO altitude is higher. Lowering the initial altitude reduces the convenience but considerably extends the range of feasibility. The reduced convenience at low altitudes is due to the fact that the Hohmann transfer requires a lower fuel quantity with the lowered the initial altitude for the same final LEO. In contrast, for the corresponding aeroassisted maneuver, even if the TPS is less consistent, the propulsive contribution remains significant.
- Figures 10-13, considered in pairs (the first two and the second two), provide a performance comparison between PICA and AVCOAT, respectively, for high initial orbits in the case of low and high aerodynamic efficiency. The comparison is in favor of the use of PICA, albeit less marked than for GEO orbits. For high orbits for a vehicle with high ε, the performance of PICA remains considerable; however, the performance of AVCOAT, although lower than PICA, is notable. In this context, it seems appropriate to comment on NASA's recent evaluation concerning the choice of material for ablative TPS of the Orion spacecraft. The final decision was between PICA and AVCOAT, and NASA ultimately chose the latter. PICA has been used to date only on a re-entry capsule of the Stardust probe, while AVCOAT has a long history of reliability (Apollo missions). Though the results of the present work provide evidence of the tangible prevalence of the PICA performance, the reliability of AVCOAT most likely outweighs these performance considerations.
- The increase in aerodynamic efficiency has less impact in the case of lower orbits (Figures 14 and 15).

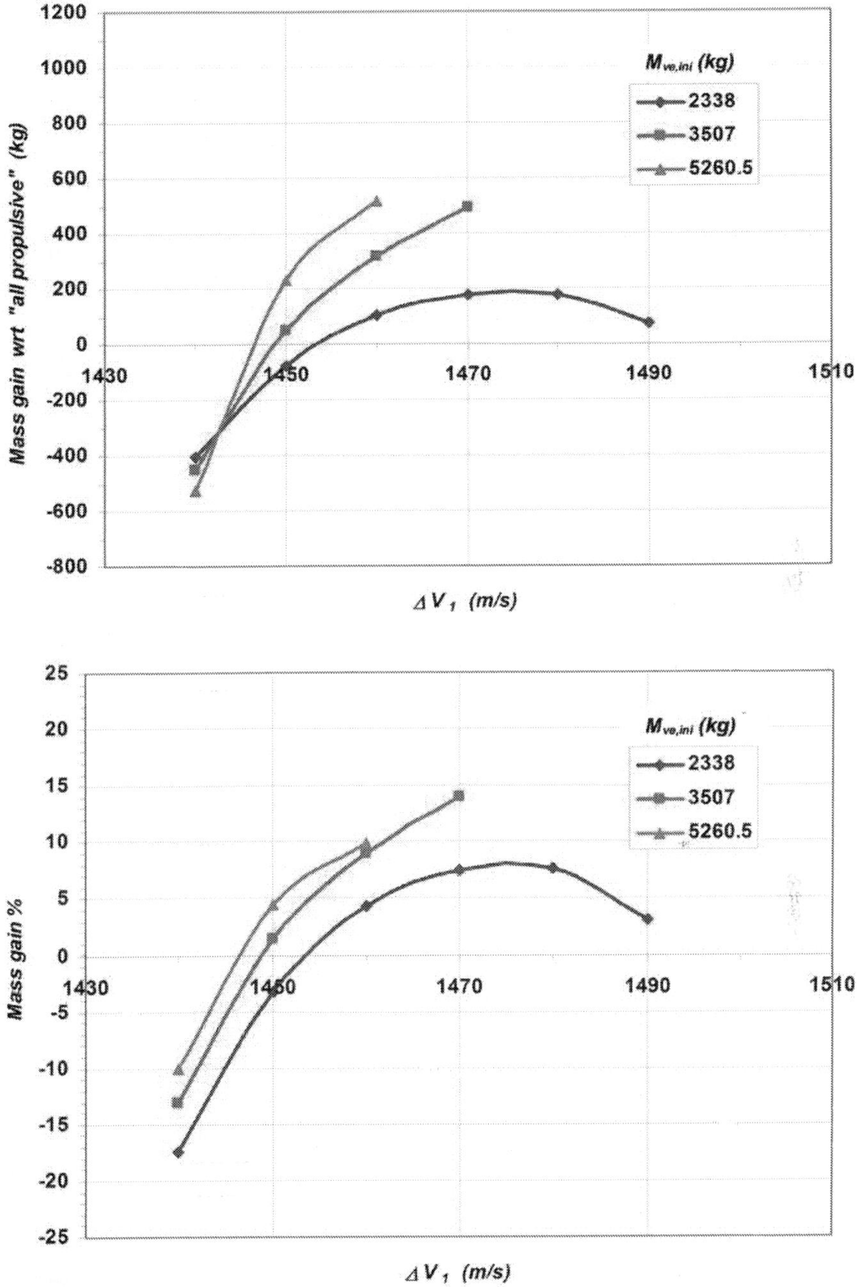

Figure 10. HEO-LEO scenario 7: HEO altitude influence; reference case for HEOH.

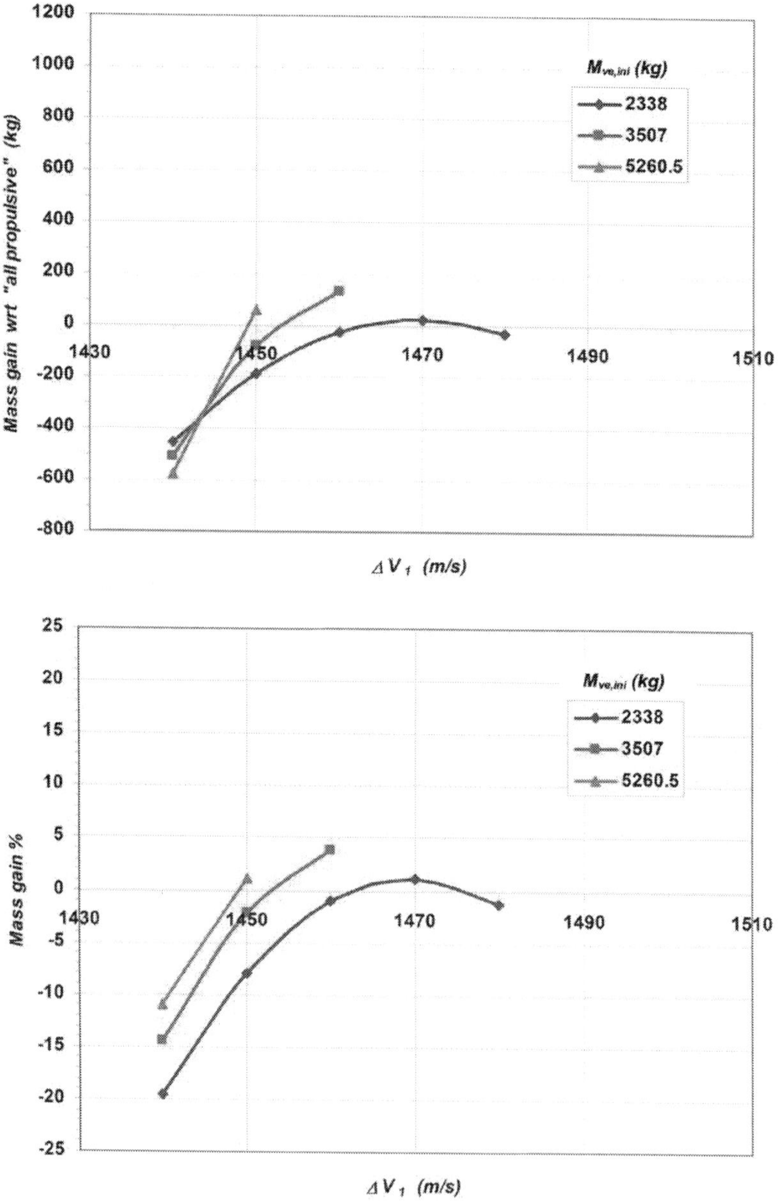

Figure 11. HEO-LEO scenario 8: material (AVCOAT) influence w.r.t. HEOH.

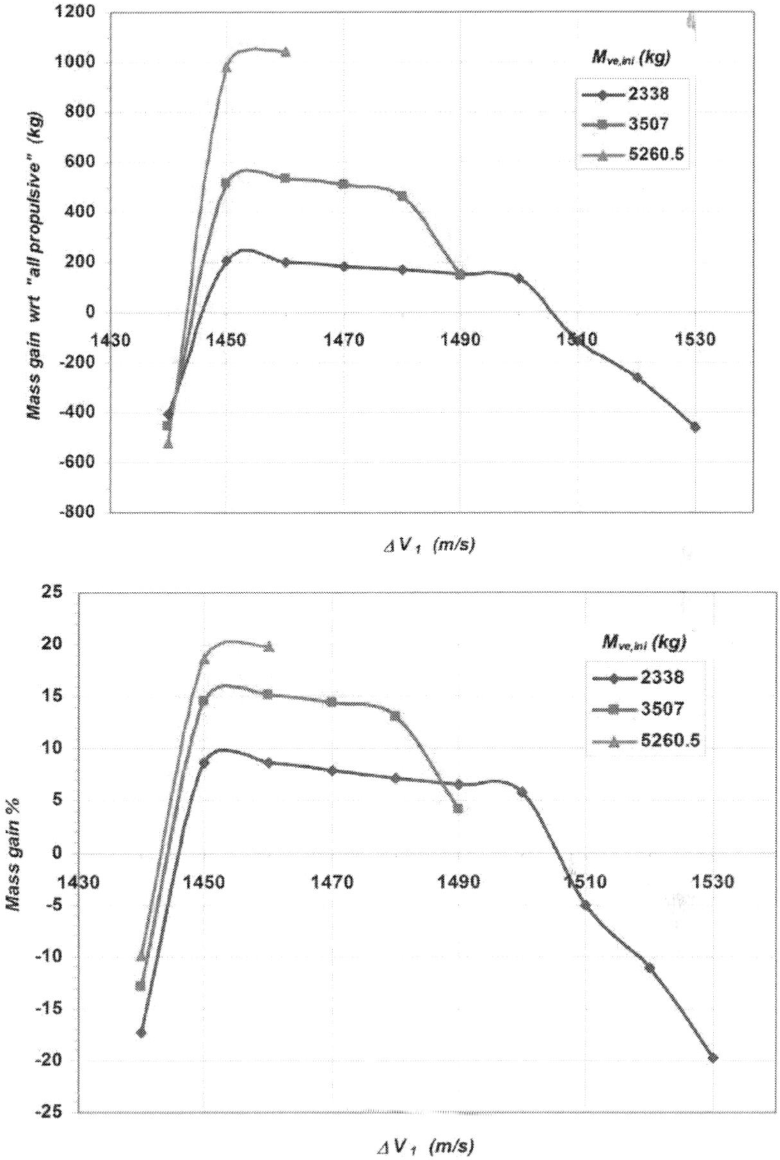

Figure 12. HEO-LEO scenario 9: e influence w.r.t. HEOH.

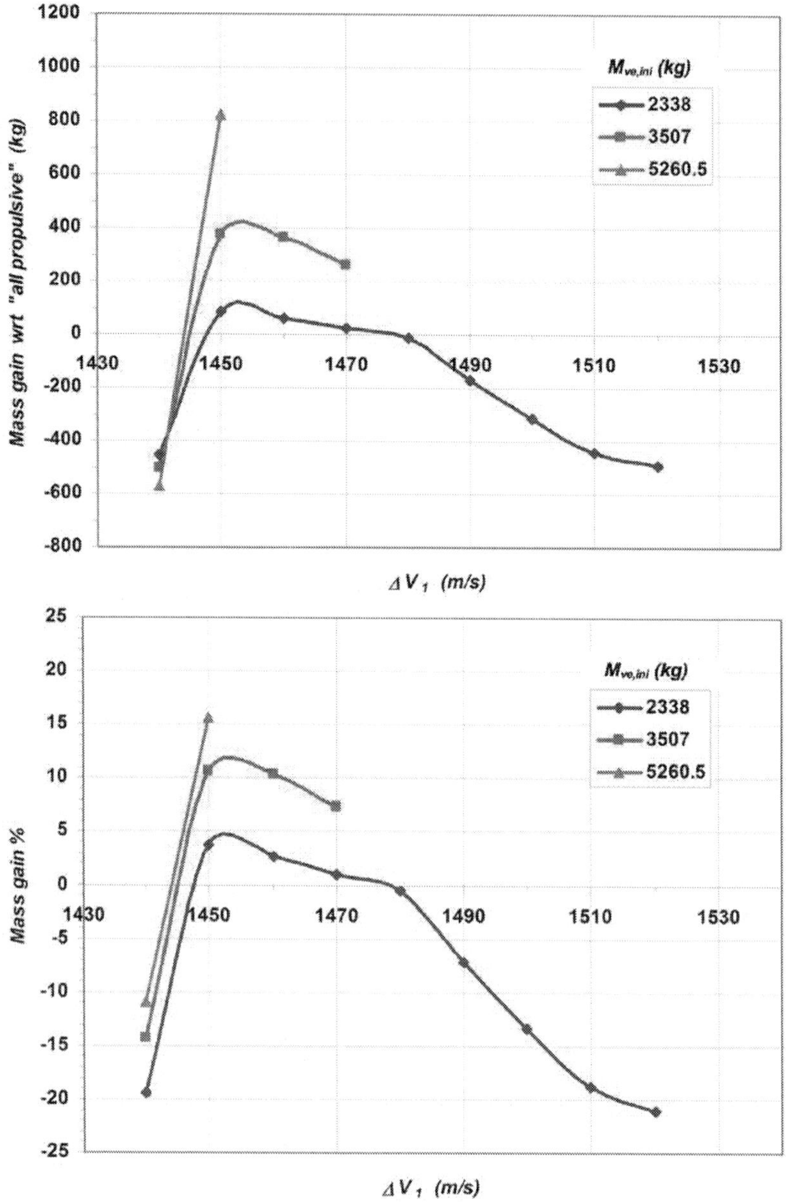

Figure 13. HEO-LEO scenario 10: Material (AVCOAT) + e influence w.r.t. HEOH.

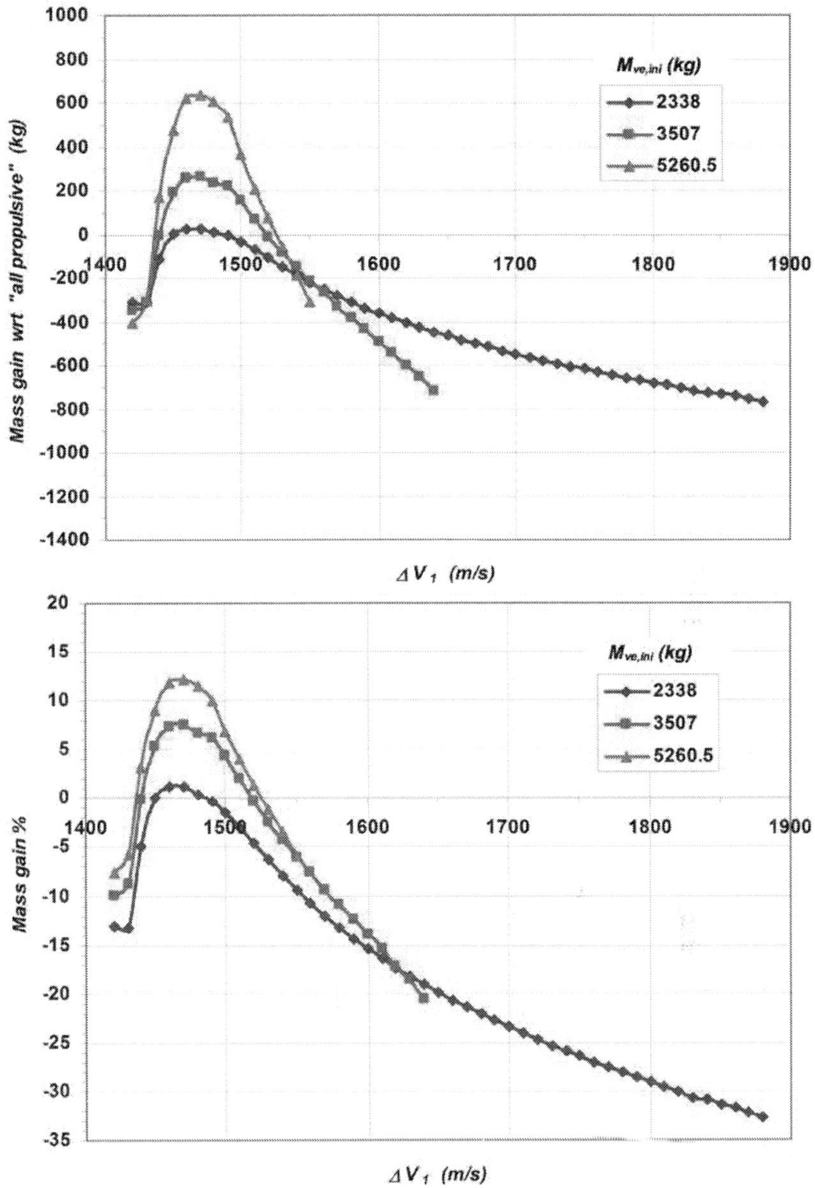

Figure 14. HEO-LEO scenario 11: HEO altitude influence; reference case for HEOL.

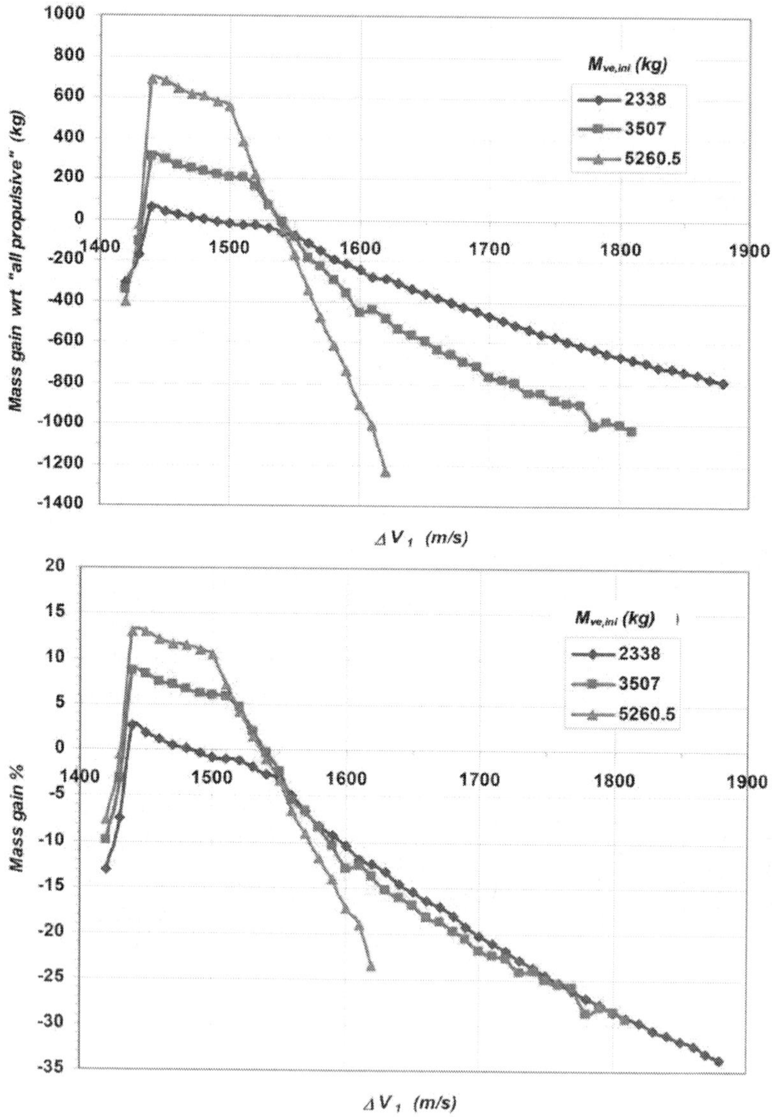

Figure 15. HEO-LEO scenario 12: e influence w.r.t. HEOL.

SUMMARY AND CONCLUSIONS

Some aeroassisted transfers between coplanar orbits were analyzed to examine the influence of the aerodynamic characteristics of a spacecraft and the characteristics of its heat shield on the maneuver's final performance.

The study was conducted with an original tool developed by the authors comprising highly representative models for thermal analysis, atmospheric flight dynamics, and an optimizer based on a genetic algorithm. The convenience of the maneuver was evaluated by comparison with the corresponding extra-atmospheric Hohmann transfer with two impulses.

The results of the case study indicated that there is a range of feasibility for the maneuver, depending on the first propulsive impulse of deorbit, within which there is, in turn, an interval in which the maneuver is more convenient than the homologous Hohmann transfer in terms of the total mass savings achieved (TPS and propellant). In particular, the results show that in terms of the convenience of the aeroassisted maneuver, the influence of the ablative material type used (low-density advantage) and of the aerodynamic efficiency is the most relevant, whereas other characteristics play a minor role. Significant benefits can be obtained from a careful thermal design, with a suitable thermal safety factor and high-performance adhesives for the bond-line.

For all cases, the highest convenience, and consequently the corresponding optimal trajectories, is obtained when it is not necessary to apply the second propulsion impulse. The most important future development for this type of analysis is increasing the number of cases analyzed in terms of materials, types of maneuvers and spacecraft characteristics.

REFERENCES

1. A. Mazzaracchio and M. Marchetti, "A Probabilistic Sizing Tool and Monte Carlo Analysis for Entry Vehicle Ablative Thermal Protection Systems," Acta Astronautica, Vol. 66, No. 5-6, 2010, pp. 821-835. doi:10.1016/j.actaastro.2009.08.033
2. A. Mazzaracchio and M. Marchetti, "Coupled Aeroassisted Orbital Plane Change Manoeuvre and Thermal Protection System Optimisation," 61st International Astronautical Congress, Prague, September 27-October 1, 2010.

3. P. Charbonneau and B. Knap, "A User'S Guide to Pikaia 1.0," Boulder, Colorado, 1995.
4. P. Charbonneau, "An Introduction to Genetic Algorithms for Numerical Optimization," Boulder, Colorado, 2002.
5. P. Charbonneau, "Release Notes for Pikaia 1.2," Boulder, Colorado, 2002.
6. C. Gogu, T. Matsumura, R. T. Haftka and A. V. Rao, "Aeroassisted Orbital Transfer Trajectory Optimization Considering Thermal Protection System Mass," Journal of Guidance, Control and Dynamics, Vol. 32, No. 3, 2009, pp. 927-938.doi:10.2514/1.37684
7. NASA Marshall Space Flight Center, "X-37 Demonstrator to Test Future Launch Technologies in Orbit and Reentry Environments," NASA Facts, May 2003, FS-2003-05- 65-MSFC.
8. Y. Y. Shi and D. H. Young, "Minimum Fuel Coplanar Aeroassisted Orbital Transfer Using Collocation and Nonlinear Programming," Flight Mechanics/Estimation Theory Symposium, NASA Goddard Space Flight Center, 1991, pp. 461-480 (SEEN92-1407005-13).
9. US Standard Atmosphere, US Government Printing Office, Washington DC, 1976.
10. T. J. Collins, W. M. Congdon, S. S. Smeltzer and K. S. Whitley, "High-Temperature Structures, Adhesives, and Advanced Thermal Protection Materials for Next-Generation Aeroshell Design," NASA Langley Research Center, 2006, Paper 2M-02-2005.

CITATION

A. Mazzaracchio and M. Marchetti, "Effect of Spacecraft Aerodynamics and Heat Shield Characteristics on Optimal Aeroassisted Transfer," Engineering, Vol. 4 No. 6, 2012, pp. 307-320. doi: 10.4236/eng.2012.46040.

Chapter 2

Spacecraft Exploration of PHOBOS and DEIMOS

Thomas C. Duxbury [a], *Alexander V. Zakharov* [b], *Harald Hoffmann* [c], *Edward A. Guinness* [d]

[a] George Mason University 6C3, 4400 University Drive, Fairfax, VA 22039, USA
[b] Institute of Space Research, Moscow, Russia
[c] Institute of Planetary Research, German Aerospace Center (DLR), Berlin, Germany
[d] Department of Earth and Planetary Sciences, Washington University, St. Louis, MO, USA

ABSTRACT

We review the previous exploration of Phobos and Deimos by spacecraft. The first close-up images of Phobos and Deimos were obtained by the Mariner 9 spacecraft in 1971, followed by much image data from the two Viking orbiters at the end of the 70s, which formed the basis for early Phobos and Deimos shape and dynamic models. The Soviet Phobos 2 spacecraft came within 100 km of landing on Phobos in 1988. Mars Global Surveyor (1996–2006) and Mars Reconnaissance Orbiter (since 2005) made close-up observations of Phobos on several occasions. Mars Express (since 2003) in its highly elliptical orbit is currently the only spacecraft to make regular Phobos encounters and has returned large volumes of science data for this satellite. Landers and rovers on the ground (Viking

Landers, Mars Pathfinder, MER rovers, MSL rover) frequently made observations of Phobos, Deimos and their transits across the solar disk.

INTRODUCTION

It was nearly 90 years after the discovery of Phobos and Deimos by Asaph Hall in 1877 at the US Naval Observatory before spacecraft were sent to Mars to explore the red planet and its moons. This section describes the many spacecraft sent to Mars that observed Phobos and Deimos and the data sets that they produced that will keep researchers busy for decades trying to answer the most basic questions of origin, evolution and current states.

The fleet of spacecraft that have observed the two Martian moons include:

Mars flyby missions:

NASA Mariners 4, 5 and 6

ESA Rosetta (still operating)

Mars orbiter missions:

NASA Mariner 9

Viking Orbiters 1 and 2

Soviet Phobos 88

NASA Mars Global Surveyor

NASA Mars Odyssey Orbiter (still operating)

ESA Mars Express (still operating)

NASA Mars Reconnaissance Orbiter (still operating)

Mars lander/rover missions:

Viking Landers 1 and 2

Mars Pathfinder

Mars Exploration Rover Spirit and Opportunity (still operating)

Mars Surface Laboratory Curiosity (still operating)

NASA spacecraft data for Phobos are available from the Planetary Data System (PDS) [1] and the National Space Science Data Center (NSSDC) [2]. Data from European Space Agency (ESA) missions can be found at [3]. Ancillary data such as trajectory, ephemerides, spacecraft attitude and clock, instrument pointing and models, etc. are available at [4]. Sample images of Phobos and Deimos taken by different missions and instruments can be found at [5] and [6].

MARINERS 4, 6 AND 7

NASA's Mariners 4, launched in 1965, and 6 and 7, launched in 1969, were Mars flyby missions. The objectives of these missions were focused on Mars with no goals to observe Phobos and Deimos. The Mariner 4 vidicon camera, with a 1.05°×1.05° field of view, obtained 21 Mars images that can be found at [7]. Other related Mariner 4 publications are: Nicks (1967), Sloan (1968), Anderson (1965), and NASA (1967).

In 1969, Mariner 6 and Mariner 7 completed the first dual mission to Mars, flying by over the equator and south polar regions and analyzing the Martian atmosphere and surface with remote sensors, as well as recording and relaying hundreds of pictures. By chance, both flew over cratered regions and missed seeing the giant northern volcanoes and the equatorial grand canyon that was discovered later. The two spacecraft each carried the Mars Wide and Narrow Angle TV Cameras, the IR Spectrometer, the Two-Channel IR Radiometer, the UV Spectrometer, the Thermal Control Flux Monitor and the Celestial Mechanics/Relativity Experiments.

It is expected that Phobos and Deimos were captured by Mariners 6 and 7 during the approach to Mars as point source images off of the Mars limb or possibly with Phobos in transit across Mars. However, the Mars approach and flyby images were not archived so one cannot go back and search for these point source images. Such observations would have extended the time span of spacecraft observations by two years, very important in determining the orbits, including secular accelerations/deceleration of the Martian moons. Publications giving more information on Mariners 6 and 7 include: Collins (1971)and NASA (1969).

MARINER 9

In 1971, Mariner 9 became the first spacecraft to orbit another planet. It carried an Infrared Radiometer (IRR), an Ultraviolet Spectrometer (UVS), an Infrared Interferometer Spectrometer (IRIS), a Television System and a Celestial Mechanics/S-Band Occultation Experiments. The objectives of these experiments were to analyze the surface and atmosphere of Mars. However bonus science was obtained when these instruments were able to view the Martian moons during close flybys.

Mariner 9 was able to map 85% of the Martian surface at a resolution of 1–2 km (with 2% mapped at a resolution of 100–300 m) in addition to gathering abundant information about the surface and atmosphere. Mariner 9 carried a wide-angle camera having a 11°×14° field of view for systematic mapping of Mars and a narrow angle camera having a 1.1°×1.4° field of view for viewing local Martian features. Seven thousand three hundred twenty nine images were taken, included the first detailed views of Olympus Mons, the solar system's largest volcano; Valles Marineris, a vast canyon system that dwarfs the Grand Canyon and was named for this spacecraft; the polar caps; and the moons Phobos and Deimos.

The Martian moons were imaged on Mars approach, against star fields, as a technical flight demonstration of a new technique for interplanetary navigation. The demonstration exceeded expectations and optical navigation has been included on many follow-on Mars, outer planet, comet and asteroid missions. Once in Mars orbit, the highly elliptic and large Mariner 9 orbit that went outside of the Deimos orbit would periodically fly near the Martian moons where images were taken anytime the range was less than 7000 km. Images as close as 1200 km were taken where a narrow angle pixel covered about 30 m.

A total of 214 images of Phobos and Deimos were obtained in greyscale from the narrow angle camera and in color using the red, green and blue filters of the wide-angle camera. These images were the first to resolve the global shapes and surface topographies showing the two Martian moons to be highly irregularly shaped and regolith-covered surface saturated with craters. The images showed Phobos and Deimos to be in synchronous rotation with Phobos having a forced rotational libration amplitude at the 1° level. A controlled drawing of the Phobos surface features was the first cartographic map of a moon of another planet with some craters given names by the IAU Nomenclature Committee. The orbital speeds of Phobos and Deimos confirmed that Phobos was speeding up as it was spiraling in toward Mars while Deimos was spiraling away from Mars due to tidal forces raised by the moons on Mars. Hints of crater chains were seen that were later resolved by Viking orbiter images to be part of the global groove network.

A summary of the entire Mariner 9 image archive, including the Phobos and Deimos images, is given by Cutts, 1974. The atlas contains a description of each image plus a thumbnail image. Digital versions of Mariner 9 images are available from the PDS at [8]. The entire image collection is also archived in photographic and digital formats at the NASA GSFC NSSDC [9].

Other websites hosting Mariner 9 data include:

UVS Data Files, Mariner Mars 1971 (MM71/M9) Ultraviolet Spectrometer (UVS) Data [10]

Experimenter Data Records files: [11]

Supplemental Experimenter Data Record files: [12]

General descriptions are at [13].

Additional information on the project and images can be found in: Steinbacher et al. (1972), Hartmann and Raper (1974), Steinbacher and Haynes (1973), Cutts (1974),Duxbury et al. (1974), Pollack et al., 1972 and Pollack et al., 1973, Gatley et al. (1974),Simmons and Hendrix (1999), and Pang et al. (1987).

VIKING ORBITERS/LANDERS

The Viking project placed two spacecraft into Mars orbit and two landers on the Martian surface in 1976. Both orbiters carried Imaging Systems, Infrared Thermal Mappers (IRTM), Mars Atmospheric Water Detectors (MAWD) and Radio Science experiments, all of which observed Phobos and Deimos during close flybys. The Lander Stereo Cameras viewed Phobos and Deimos from the surface of Mars. The Viking project elevated Phobos and Deimos exploration as a high science objective during its extended mission phase and targeted many close flybys of the Mars moons. Additionally, the shadow of Phobos passing over Viking Lander 1 was observed simultaneously three times from the lander and Viking

Orbiter 1. The Viking orbiter/lander datasets stood as the most exhaustive data on the Martian moons for almost 40 years, only recently being exceeded by the ESA Mars Express mission.

Viking had many new discoveries related to Phobos and Deimos:

The global network of grooves;

Complete global coverage of the entire surfaces with significant coverage in color, UV and IR;

The first digital terrain model (DTM) of another planet's moon (Phobos) and the first global image mosaic (also Phobos);

The first mass determinations of the two moons showing that they were anomalous to Mars and asteroids by having densities of less than 2 g/cm^3; Phobos and Deimos surfaces were covered by fine-grained regoliths, tens of meters or thicker; and

Phobos had a forced rotational libration of about 1° amplitude and that this amplitude was consistent with Phobos having a homogeneous interior (low density material or loosely compacted Mars primordial disk material or crater ejecta), covered by regolith.

All Viking Orbiter and Lander data of Phobos and Deimos are archived at NASA's PDS and NSSDC facilities: [14] (Viking Orbiter) and [15] (Viking Lander).

A photo journal of the close flyby images of Phobos is given in Duxbury et al. (1984). A detailed report of the IR results can be found in Lunine et al. (1982).

PHOBOS 88

The Soviet Phobos 2 spacecraft went into Mars orbit in 1988 for three months and came within 100 km of landing on Phobos. Phobos 2 carried many orbiter and lander instrument. The orbiting remote sensing instruments that observed Phobos and Deimos included: the Automatic Space Plasma Experiment with Rotating Analyzer (ASPERA), the Videospectrometric System (VSK), the Infrared Mapping Spectrometer (ISM), the Infrared Radiometer/Spectrometer (KFRM), and the Scanning Infrared Radiometer (Termoscan).

The VSK included two wide-angle cameras (red and blue channels) and a high-resolution narrow-angle camera to yield visual and color images of

Phobos. VSK also captured Deimos as a point source with Jupiter and the bright star Aldebaran.

The science and associated ancillary engineering data were archived by the Soviet Academy of Sciences Space Research Institute (IKI) and the NASA Planetary Data System (PDS) and NASA Ancillary Information Facility (NAIF) under the US/Soviet Joint Working Group for Solar System Exploration.

The ISM data archive is at [16]. The KRFM, Termoscan, and VSK data are available at [17]. The ancillary engineering data are archived as NAIF SPICE kernels at [18].

Data returned from the Phobos-1 and -2 missions		
Experiment characteristics	Available data	References
Video spectrometric system (VSK) Fregat	37 TV images of Phobos (and Phobos against Mars) from distance of 190–1100 km	Avanesov et al. (1989)
Multichannel instrument KRFM	Phobos observation from a height of 190 km on 25.03.1989	Ksanfomality et al. (1989)
–Radiometer 6–50 μm –Photometer 320–600 nm	Observation period 420 s	
	Relative speed 43 m/s	
	1st track at the surface	
	Latitude: –4.5°	
	Longitude: 140–260° West	
Thermal and special properties of the Phobos regolith	2nd track	
	Latitude: 2.7°	
	Longitude: 196–229° West	
Imaging spectrometer ISM	Mars observation:	Bibring et al. (1989)
–1.65–3.16 μm and –0.76–1.54 μm	From <2000 km	
Surface mineralogy of Mars and Phobos, composition of the Martian atmosphere	40,000 spectra of Mars and Phobos	
Themoscan, thermal imaging of the surface of Mars Radiometer	The first session of measurement – on 11.02.1989 (intermediate elliptical orbit, altitude during observation 1150–5200 km, resolution ~300 m)	Selivanov et al. (1989)
•8.5–12 μm and •0.5–0.95 μm	The second session of measurement – on 01.03.1989 (circular orbit,	

	altitude 6300 km)	
Gamma-ray Spectrometer 100×100 mm² CsI(Tl) crystal 0.3–10.0 MeV Elemental composition of the Martian soil	Measurements of γ-radiation from Mars during passing the perigee at the first four elliptical orbits, the altitude of the perigee ~870 km (1, 5, 8 and 11 February 1989).	Surkov et al. (1989)
APEX experiment Temporal variations and spectral characteristics of cosmic γ-ray emissions, γ-ray albedo of the Martian surface. Range – 60 keV–9 MeV 100×100 mm² CsI(Tl) crystal	Measurements of γ-radiation from Mars during passing the perigee at the first four elliptical orbits, the altitude of the perigee ~870 km (1, 5, 8 and 11 February 1989).	d' Uston et al. (1989)
Solar occultation experiment	32 occultations in the period between 08.02.1989 and 17.02.1989	Blamont et al. (1989),Krasnopolsky et al. (1989)
Vertical profile of the Martian atmosphere –Spectrograph: spectral range 215–328 nm (512 elements) –Fabry-Perot interferometer 760 nm (Δ 0.2 nm, 80 elements) and 936 nm (Δ 0.3 nm, 80 elements) –Grating spectrograph (5290–5370 cm^{-2} 2707–2740 cm^{-2} 24 elements each)		
MAGMA (magnetic fields near Mars) and FGMM (flux-gate magnetometer Mars) Range of both magnetometers +/−100 nT	Measurements of the magnetic field during operation of the spacecraft at the elliptical and the circular orbits around Mars	Riedler et al. (1989)
PWS, plasma wave system Two sensors: dipole antenna and Langmuir probe	Measurements of the electromagnetic and electrostatic waves during operation of the spacecraft at the elliptical and the circular orbits around Mars	Grad et al. (1989)
ASPERA, automatic space plasma experiment with a rotation analyzer Measurements energy, and angular distribution of ions of charge e with energies 0.5 eV e^{-1}–24 keV e^{-1} and electrons with energies 1 eV–50 keV	Measurements of the energy spectra of ions and electrons and their angular distribution during operation of the spacecraft at the elliptical and the circular orbits around Mars	Lundin et al. (1989)
TAUS experiment Ions (H^+,	Measurements of the energy	Rosenbauer et al.

He^+ and "heavy ions") distribution and dynamics in the Martian environment	spectra of ions during operation of the spacecraft at the elliptical and the circular orbits around Mars	(1989)
Energy range 30 V–6 kV		
HARP spectrometer.	Measurements of the energy spectra of ions during operation of the spacecraft at the elliptical and the circular orbits around Mars	Shutte et al. (1989)
Hyperbolic analyzer in the retarding potential mode		
Energy range 0.2–800 eV		
SLED	Measurements of the energetic electrons and ions during operation of the spacecraft during the cruise phase and at the elliptical and the circular orbits around Mars	Afonin et al. (1989),McKenna-Lawlor et al. (1991)
Energy spectra of charge particles in the energy range from 30 keV to a few MeV		
LET, low energy telescope	Measurements during the 9-months interval from July 1988 to March 1989	Marsden et al. (1991)
Spectra range 1–75 MeV		
LILAS	Measurements in the period from July 1988 to March 1989	Barat et al. (1991)
Investigation of cosmic gamma-rays bursts in the range 3–1000 keV	Recorded 72 cosmic rays bursts during 200 effective days of operations	
RF-15, broadband photometry of the solar soft X-rays 294 and 4–8 keV	Measurements in the period between 14.07.1988 and 17.03.1989	Valnicek et al. (1991a)
SUFR, solar ultraviolet radiometer	Measurements on 8–11 March and 23 March 1989	Kazachevskaya et al. (1991)
Record solar emission flux atλ<130 nm		
IPHIR	24 measurements	Bruns and Shumoko (1991), Bruns et al. (1990)
–three-channel sun-photometer (SPM) – 335, 500 and 862 nm –two-axis solar sensor (TASS) –data processing unit		
TEREK experiment,	Observations:	Valnicek et al. (1991b)
Onboard the Phobos-1 spacecraft.	23.07.1988	
Imaging of the whole solar disk in X (0.5–2.5 nm), E.U.V. (17–18, 30.4 nm) and optical coronograph (400–600 nm)	26.07.1988	
	27.07.1988	
	Altogether 448 images were received	
Celestial mechanic experiment	Measurements during orbiting around Mars	Kolyuka et al. (1991)
Ground based measurements		

MARS PATHFINDER

The 1996 Mars Pathfinder Rover made multispectral measurements of both Phobos and Deimos from the surface of Mars using its Imager for Mars Pathfinder (IMP) camera (Thomas et al., 1999 and Murchie et al., 1999). Mars Pathfinder image data are available from the PDS at [19].

MGS

The Mars Global Surveyor (MGS) spacecraft went into Mars orbit in 1997 and performed flawlessly for 10 years. Its scientific remote sensing payload that observed Phobos, Deimos and the shadow of Phobos included: the Mars Orbiter Camera (MOC), the Thermal Emission Spectrometer (TES), and the Mars Orbiter Later Altimeter (MOLA). MGS was a polar orbiter flying about 400 km above Mars giving no opportunity for any close flybys of the Martian moons during this mission phase. However as MGS initially went into Mars orbit and was using aerobraking to obtain the low circular orbit, close flybys of Phobos were targeted and MOC, TES, and MOLA data were obtained from Phobos. Once in the low, circular orbit, a few images of Deimos and the shadow of Phobos were taken (Bills et al., 2005). MSSS MOC images of Phobos, Deimos and shadows (Fig. 1) can be found at [20] or [21]. Other MGS data including TES and MOLA data are archived at [22].

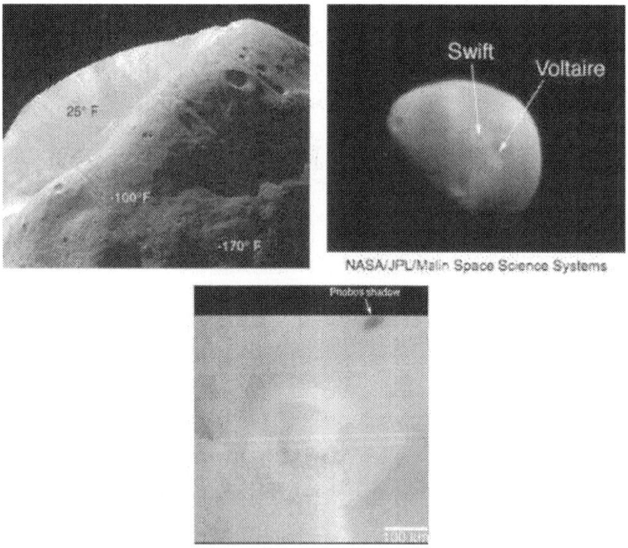

Figure. 1. NASA MSSS MOC images of the Phobos (upper left) with temperatures from TES, Deimos (upper right) and the Phobos shadow (bottom).

ODYSSEY

The Mars 2001 Odyssey spacecraft went into a low, circular Mars orbit in October 2001 and did not observe Phobos or Deimos. However, the Thermal Emission Imaging System camera (THEMIS) observed the shadow of Phobos in February 2012. The Mars 2001 Odyssey orbiting spacecraft is still operating.

MRO

The Mars Reconnaissance Orbiter went into its low, near-circular orbit in August 2006. During approach, Phobos and Deimos were observed as point sources against star fields by the Navigation Camera. This camera had a 500 mm focal length, 12 µm×12 µm pixels in a 1020×1024 CCD. Over 500 Deimos star images and over 100 Phobos star images were taken. Reduced astrometric data can be located at [23] and had positional accuracies at the km level.

While in orbit, MRO occasionally imaged Phobos (Fig. 2) and Deimos with its High Resolution Imaging Science Experiment (HiRISE) and the hyper-spectral infrared Compact Reconnaissance Imaging Spectrometer for Mars (CRISM). The NASA MRO orbiting spacecraft is still operating and expected to periodically add new observations of Phobos and Deimos. The HiRISE observations are described in Thomas et al. (2010)and can be found at [24]. The CRISM observations are described in Murchie et al. (2008). CRISM data are archived by the PDS at [25].

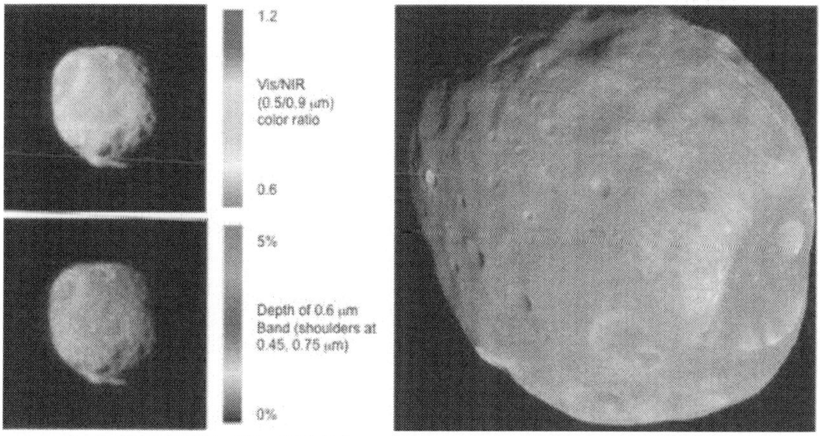

Figure. 2. NASA MRO APL CRISM and University of Arizona HiRISE images of Phobos.

MER ROVERS

Multiple images of Phobos and Deimos transits across the sun (Fig. 3) were taken by the two MER Rovers, Spirit and Opportunity, and continue to be taken by Opportunity using the Pancam system [26]. The complete archive of MER images within the PDS at [27] and [28], also see [29].

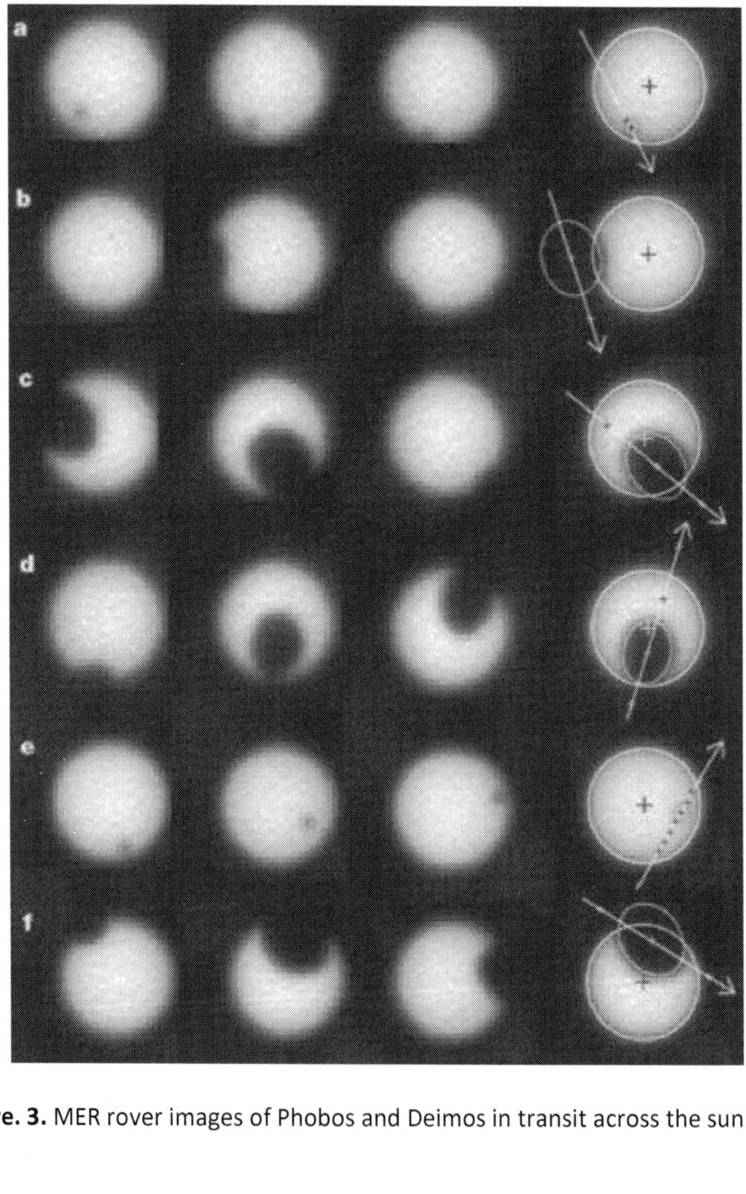

Figure. 3. MER rover images of Phobos and Deimos in transit across the sun.

MEX

Chapter XI is entirely devoted to the ESA Mars Express Exploration of Phobos and Deimos. A short history of this exploration is included here for completeness of this chapter so that it can stand alone. The European Mars Express mission arrived at Mars on 25 December 2003. Unfortunately, the lander Beagle 2 was lost after atmospheric entry and descent. The core of the mission, however, the Mars Express orbiter was inserted successfully in an elliptical orbit around Mars having an inclination of 86.6°. With a period of 7.6 h, a periapsis height of about 260 km and a semi-major axis close to 9300 km, Mars Express is routinely going beyond Phobos' orbit. Despite several orbit changes to adjust for scientific needs for Mars observations, the general characteristics of the Mars Express orbit remained similar. The satellite approaches Phobos about every 500 orbits at distances well below 1000 km (Fig. 4). With orbit number 11,453 on 31 December 2012, the Mars Express has completed its 3rd mission extension and continues successfully its scientific operations after nine years around Mars. The orbit now has a period of 7 h with its periapsis at 350 km and a semi-major axis of 8850 km. Mars Express performed small orbit phasing maneuvers since 2008 (after orbit 6000) to increase the number of very close flybys around 100 km to Phobos and below.

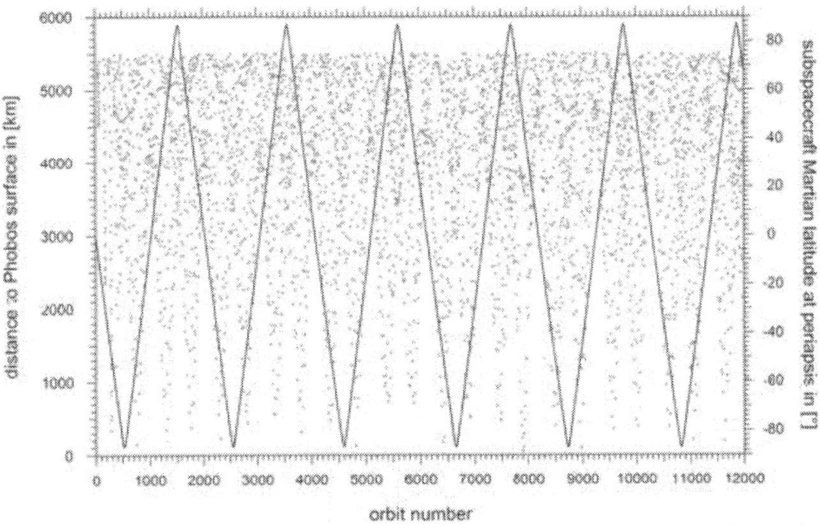

Figure. 4. Encounters of Mars Express with Phobos at distances to the surface below 5500 km (dots) compared with the latitudinal drift of the periapsis (solid line) since Mars Express orbit insertion.

Mars Express carries seven experiments with the High Resolution Stereo Camera (HRSC), the imaging spectrometer Observatoire pour la Minéralogie, l'Eau, les Glaces et l'Activité (OMEGA), the Planetary Fourier Spectrometer (PFS), the UV-IR spectrometer Spectroscopy for the Investigation of the Characteristics of the Atmosphere of Mars (SPICAM), the radar instrument Mars Advanced Radar for Subsurface and Ionosphere Sounding (MARSIS), the plasma package Analyzer of Space Plasmas and EneRgetic Atoms (ASPERA-3), and the Mars Radio Science experiment (MaRS). All instruments contributed to the Mars Express investigation of Phobos and a detailed summary of the observations and achievements is given in the accompanying paper by Witasse et al. (2013). In addition, Mars Express is equipped with a technical camera system, the Video Monitoring Camera (VMC), designed for the documentation of the lander release, which is used since 2008 for monitoring purposes and public outreach activities. All Mars Express observation data are and will be archived in ESA's Planetary Science Archive (PSA) [30] and NASA's Planetary Data System (PDS) [31].

OMEGA is an imaging spectrometer from 0.38–5.1 µm comprising three wavelength channels. In the visible, it is operated as a pushbroom scanner while in the near- and shortwave-infrared OMEGA works as a whiskbroom scanner. OMEGA used close flybys for Phobos observations to decipher its surface composition.

PFS is a point spectrometer from 1.2 µm to 45 µm optimized for atmospheric studies at Mars. With its FOV of 1.7° in the short wavelength range and of 2.8° in the long wavelength range, PFS acquires spectra of Phobos at distances well below 1000 km.

SPICAM is a dual spectrometer for investigations of the Martian atmosphere. It consists of an UV channel covering the wavelength range of 118–320 nm and a near-infrared channel from 1.0 µm to 1.7 µm. SPICAM performs Phobos measurements to better understand its composition mostly together with the HRSC camera and is operating also during far-distant observations.

MARSIS is an active sounding radar instrument operating in the wavelength range of 0.1–5.5 MHz. Its main interest with respect to Phobos is to decipher the internal structure of the Martian moon. Technical reasons constrain MARSIS observations to distances between about 100 km and 500 km.

MaRS makes use of the Mars Express communication system for radio science investigations around Mars including gravity, bistatic radar, ionospheric sounding and investigations of the solar corona during conjunctions. Measurements at Phobos are performed by pointing the antenna towards Earth and are aiming to better constrain its mass and internal structure. First results have demonstrated that very close flybys like those planned for the next years with flyby distances below 50 km are needed for further improvements.

ASPERA-3 finally comprises four different sensors to investigate the interactions of the solar wind with the Martian atmosphere and to investigate the plasma and neutral gas environment of Mars including the influence by the Martian moons. Measurements in the vicinity of Phobos have been performed whenever feasible.

The HRSC is a multiple-line single-optics pushbroom scanner with five panchromatic channels for nadir high-resolution (40 m/pixel at 1000 km distance) and quintuple stereo imaging as well as four multispectral channels for color imaging. An additional channel with separate optics and a frame detector, the Super Resolution Channel (SRC), yields a four times better spatial resolution. Since the end of 2012, HRSC was switched on in 3513 orbits and has acquired more than 39,000 images including 8474 SRC frames until now. One hundred ninety nine of these orbits were used by HRSC to acquire imaging sequences of Phobos. The pushbroom scanner was mainly applied to obtain stereo and/or color imagery at distances preferably ≤1500 km (Fig. 5). The SRC performed additional far-distant astrometric observations ≤5500 km together with the nadir channel. Achievements of the HRSC are discussed in several accompanying papers of this issue.

Figure. 5. Imaging by the HRSC pushbroom scanner: (a) Phobos above the Martian limb in orbit 6906 at 530 km distance with the P1 panchromatic channel. (b) Phobos above the Martian surface in orbit 7982 at a distance of 1382 km.

On 5 November 2009, Mars Express was pointed towards Phobos to obtain a video sequence with the SRC showing a Deimos occultation by Phobos (Fig. 6) that can be found at [32, 33, 34, 35]. A Jupiter occultation by Phobos was observed by SRC on 1 June 2011[36, 37]. The HRSC also observed the shadow of Phobos on the Martian surface in ten imaging sequences. Mars Express was specifically pointed off nadir to capture the passage of Phobos' shadow on nine of the ten sequences. One image sequence occurred in normal nadir attitude. A similar search for detecting the penumbra of Deimos by HRSC was not successful.

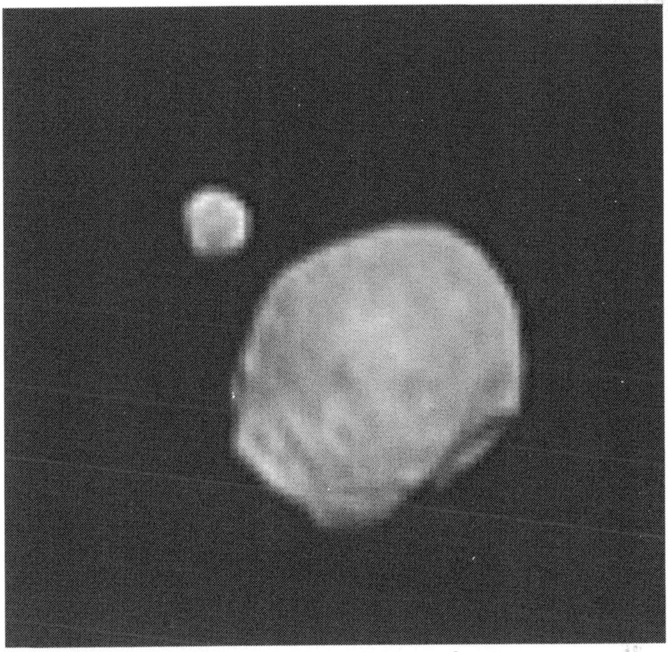

Figure. 6. SRC image of Phobos partial occultation of Deimos sequence.

The second Martian moon Deimos is well outside of the Mars Express orbit. The closest approach to Deimos during nominal mission was well above 9500 km. Observations are restricted to the Mars facing side and spatial resolutions ≤100 m/pixel are barely achieved by SRC. Main emphasis for investigating Deimos was given to refine its orbital parameters and to better constrain its surface composition. HRSC has obtained imaging sequences during 68 orbits with its nadir channel together with the SRC channel. SPICAM was operating together with HRSC for most of the Deimos observations. Rather recently, OMEGA has also increased its number of Deimos observations with its visible channel.

Searches for rings and moons of Mars were carried out using both the HRSC Nadir and Super Resolution channels as the MEX spacecraft passed through the Mars equatorial plane on five different occasions. Image sequences were obtained at both low (backward scattered light) and high (forward scattered light) solar phase angles. To date, nothing has been observed; however additional attempts will be made using the experience gained from the existing sequences.

Mars Express data are archived within the ESA's Planetary Science Archive (PSA) at [38] and with the PDS at [39]. The MEX observation of Phobos and Deimos will continue beyond what is reported here as the spacecraft is still active in its operations at Mars.

ROSETTA

The ESA spacecraft Rosetta flew by Mars for a gravity assist before continuing its journey to a comet. On 24 February 2007 before closest approach, the Optical, Spectroscopic and Infrared Remote Imaging System (OSIRIS) observed Phobos and its shadow in transit of Mars. A video animation is available at [40]. More details about the OSIRIS NAC multispectral observations of Phobos before and after closest approach are described by Pajola et al. (2012).

MSL

MSL Curiosity is observing Phobos/Deimos solar transits (Fig. 7), mutual encounters and encounters with Jupiter and Saturn to support orbit evolution and Mars interior studies. MSL data can be found from the PDS at [41].

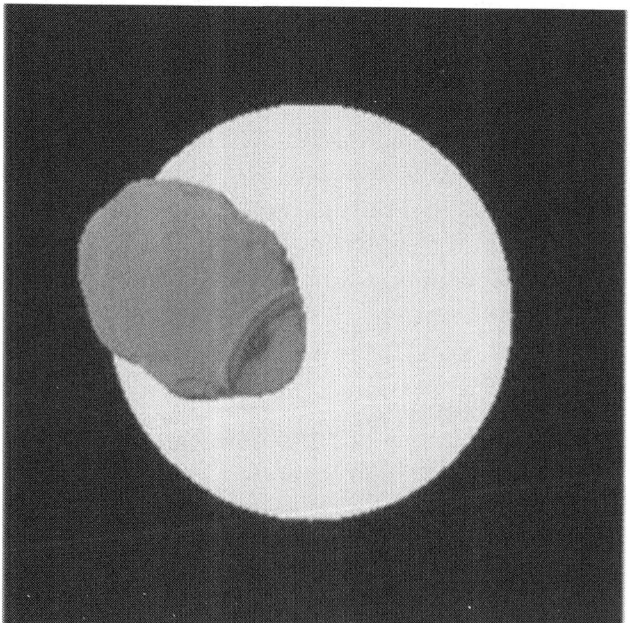

Figure. 7. Predicted solar transit by Phobos as viewed from MSL curiosity.

ACKNOWLEDGMENT

Appreciation is given to the NASA GSFC National Space Science Data Center (NSSDC)for providing information on the early Mars Missions and ISSI for supporting meetings and logistics that lead to this chapter. Work supporting this chapter was partially funded by the NASA Mars Express Participating Scientist Program.

REFERENCES

1. Afonin, V., McKenna-Lawlor, S., Gringauz, K., Kecskemety, K., Keppler, E., Kirsch, E., Richter, A., OO Sullivan, D., Somogyi, A., Thompson, A., Varga, A., Witte, M., 1989. Energetic ions in the close environment of Mars and particle shadowing by the planet. Nature 341 (6243), 616–618
2. Anderson, H.R., 1965. Mariner 4 measurements near Mars, initial results, spacecraft description and encounter sequence. Science 149 (3689), 1226–1228.
3. Avanesov, G.A., Bonev, B.I., Kempe, F., Bazilevsky, A.T., Boycheva, V., Chilkov, K.N., Danz, M., Dimitrov, D., Duxbury, T., Gromatikov, P., Halmann, D., Head, J., Heifets, V.N., Kolev, V., Kostenko, V.I., Kottsov, V.A., Krasavtsev, V.M., Krasikov, V.A., Krumov, A., Kuzmin, A.A., Losev, K.D., Lumme, K., Mishev, D.N., Möhlmann, D., Muinonen, K., Muravaev, V.M., Murchie, S., Murray, B., Neumann, W., Paul, L., Petkov, D., Petuchova, I., Pössel, W., Rebel, B., Shkuratov, Yu G., Simeonov, S., Smith, B., Totev, A., Uzunov, Y., Fedotov, V.P., Weide, G.G., Zapfe, H., Zhukov, B.S., Ziman, Y.L., 1989. Television observations of Phobos. Nature 341 (6243), 585–587.
4. Barat, C., Atteia, J.L., Jourdan, E., Vedrenne, G., Sunyaev, R., Blinov, N., Kuznetsov, A., Lvov, M., Terekhov, O., 1991. On the absorption features in cosmic gamma-ray burst spectra recorded by the LILAS experiment. Planet. Space Sci. 39 (1/2), 67–71.
5. Bibring, J.P., Combes, M., Langevin, Y., Soufflot, A., Cara, C., Drossart, P., Encrenaz, T., Erard, S., Forni, O., Gondet, B., Ksanfomality, I., Lellouch, E., Masson, P., Moroz, V., Rocard, F., Rosenqvist, J., Sotin, C., 1989. Results from the ISM experiment. Nature 341 (6243), 591–593.
6. Bills, B.G., Neumann, G.A., Smith, D.E., Zuber, M.T., 2005. Improved estimate of tidal dissipation within Mars from MOLA observations of the shadow of Phobos. J. Geophys. Res. 110, E07004, http://dx.doi.org/10.1029/2004JE002376.
7. Blamont, J.E., Chassefiere, E., Goutail, J.P., Mege, B., Nunes-Pinharanda, M., Souchon, G., Krasnopolsky, V.A., Krysko, A.A., Moroz, V.I., 1989. Vertical profiles of dust and ozone in the martian atmosphere deduced from solar occulation measurements. Nature 341 (6243), 600–603

8. Bruns, A.V., Bonnet, R., Delaboudinier, J.P., Frohlich, C., Shumko, S.M., 1990. Results of solar brightness Oscillations—PHOBOS-2 observations. 16 (2), 140
9. Bruns, A.V., Shumoko, S.M., 1991. Short term variability of the power spectrum of 5-min oscillations of the Sun. Planet. Space Sci. 39 (1/2), 41–47.
10. Collins, S.A., 1971. Mariner 6 and 7 pictures of Mars. NASA, SP-263, Washington, D.C.
11. Cutts, J., 1974. Mariner 9 Picture Catalog: vol. 1, Jet Propulsion Laboratory Report. TR 33-585, 422 p.
12. Duxbury, T.C., Born, G.H., Jerath, N., 1974. Viewing Phobos and Deimos for Navigating Mariner 9. J. Spacecr. Rockets 11 (4), 215–222, http://dx.doi.org/10.2514/3.62046
13. Duxbury, T.C., Callahan, J.D., Ocampo, A.C., 1984. Phobos: Close Encounter Imaging from the Viking Orbiters. US National Aeronautics and Space Administration, Washington D.C., USA
14. Gatley, I., Kieffer, H., Miner, E., Neugebauer, G., 1974. Infrared Observations of PHOBOS from Mariner 9. Astrophys. J. 190, 497–508, http://dx.doi.org/10.1086/ 152902.
15. Grad, R., Pedersen, A., Klimov, S., Savin, S., Skalsky, A., Trotignon, J.G., Kennel, C., 1989. First measurements of plasma waves near Mars. Nature 341 (6243), 607–609.
16. Hartmann, W.K., Raper, O., 1974. The New Mars: The Discoveries of Mariner 9. NASA, SP-337, Washington, D.C
17. Kazachevskaya, T.V., Bukusova, L.L., Gonyukh, D.A., Lomovsky, A.I., Tsygelnitsky, Y. N., 1991. E.u.v. observations of solar flares from Mars. Planet. Space Sci. 39 (1/2), 39–41.
18. Kolyuka, Y.F., Kudryavtsev, S.M., Tarasov, V.P., Tikhonov, V.F., Ivanov, N.M., Polyakov, V.S., Potchukaev, V.N., Parkov, O.V., Sukhanov, K.G., Akim, E.I., Stepanians, V.A., Nasirov, R.R., 1991. International project Phobos. Experiment celestial mechanics. Planet. Space Sci. 39 (1/2), 349–355.
19. Krasnopolsky, V.A., Moroz, V.I., Krysko, A.A., Korablev, O.I., Zhegulev, V.S., Grigoriev, A.V., Tkachuk, A.Y., Parshev, V.A., Blamont, J.E., Goutail, J.P., 1989. Solar occultation spectroscopic measuremens of the martian atmosphere at 1.9 and 3.7 mm. Nature 341 (6243), 603–604.
20. Ksanfomality, L.V., Moroz, V.I., Bibring, J.P., Combes, M., Soufflot, A., Ganpantzerova, O.F., Goroshkova, N.V., Zharkov, A.V., Nikitin, G.E., Petrova, E.V., 1989. Spatial variations in thermal and albedo properties of the surface of Phobos. Nature 341 (6243), 588–591.
21. Lundin, R., Zakharov, A., Pellinen, R., Borg, H., Hultqvist, B., Pissarenko, N., Dubinin, E.M., Barabash, S.W., Liede, I., Koskinen, H., 1989. First measurements of the ionospheric plasma escape from Mars. Nature 341 (6243), 609–612.

22. Lunine, J.I., Neugebauer, G., Jakosky, B., 1982. Infrared observations of Phobos and Deimos from Viking. J. Geophys. Res. 87, 10297.
23. Marsden, R.G., Wenzel, K.P., Afonin, V.V., Gringauz, K., Witte, M., Richter, A.K., Erdös, G., Somogyi, A.J., Varga, A., Varhalmi, L., 1991. Energetic participle composition measurements from Phobos 2: results of the LET experiment. Planet. Space Sci. 39 (1/2), 57–66.
24. McKenna-Lawlor, S.M.P., Afonin, V.V., Gringauz, K.I., Keppler, E., Kirsch, E., Richter, A.K., Witte, M., O0 Sullivan, D., Thompson, A., Kecskemety, K., Somogyi, A., Szabo, L., Varga, A., 1991. Interplanetary variability recorded by the SLED instrument aboard the Phobos spacecraft during that period of solarcycle 22 characterized by a transition from solar minimum-to solar maximum dominated conditions. Planet. Space Sci. 39 (1/2), 47–57.
25. Murchie, S., Thomas, N., Britt, D., Herkenhoff, K., Bell , J.F., 1999. Mars Pathfinder spectral measurements of Phobos and Deimos: comparison with previous data. J. Geophys. Res. 104 (E4), 9069–9079, http://dx.doi.org/10.1029/98JE02248.
26. Murchie, S.L., Choo, T., Humm, D., Rivkin, A., Bibring, J.-P., Langevin, Y., Gondet, B., Roush, T., Duxbury, T., 2008. MRO/CRISM observations of Phobos and Deimos. In: Proceedings of the 39th Lunar and Planetary Science Conference, abstract #1434
27. NASA, 1967. Mariner-Mars Final Project Report 1964. SP-139, Washington, D.C.
28. NASA, 1969. Mariner-Mars 1969 – a Preliminary Report: Scientific Experiments. SP- 225, Washington, D.C
29. Nicks, O.W., 1967. Review of the Mariner 4 results. Moon Planets, 150–185
30. Pajola, M., Lazzarin, M., Bertini, I., Marzari, F., Turrini, D., Magrin, S., La Forgia, F., Thomas, N., Küppers, M., Moissl, R., Ferri, F., Barbieri, C., Rickman, H., Sierks, H., the OSIRIS Team, 2012. Spectrophotometric investigation of Phobos with the Rosetta OSIRIS-NAC camera and implications for its collisional capture. Mon. Not. R. Astron. Soc. 427 (4), 3230–3243, http://dx.doi.org/10.1111/j.1365- 2966.2012.22026.x.
31. Pang, K.D., Slavin, J.A., Franklin, B.O., 1987. Mariner 9 Far UV observations of Phobos and Deimos: search for evidence of outgassing. Bull. Am. Astron. Soc. 19, 835.
32. Pollack, J.B., Veverka, J., Noland, M., Sagan, C., Hartmann, W.K., Duxbury, T.C., Born, G.H., Milton, D.J., Smith, B.A., 1972. Mariner 9 television observations of Phobos and Deimos. Icarus 17 (2), 394–407
33. Pollack, J.B., Veverka, J., Noland, M., Sagan, C., Duxbury, T.C., Acton , C.H., Born, G.H., Hartmann, W.K., Smith, B.A., 1973. Mariner 9 television observations of Phobos and Deimos, 2. J. Geophys. Res. 78 (20), 4313–4326, http://dx.doi.org/10.1029/ JB078i020p04313.
34. Riedler, W., Möhlmann, D., Oraevsky, V.N., Schwingenschuh, K., Yeroshenko, Y., Rustenbach, J., Aydogar, O., Berghofer, G., Lichtenegger, H., Delva, M., Schelch, G., Pirsch, K., Fremuth, G., Steller, M., Arnold, H., Raditsch, T.,

Auster, U., Fornacon, K.H., Schenk, H.J., Michaelis, H., Motschmann, U., Roatsch, T., Sauer, K., Schröter, R., Kurths, J., Lenners, D., Linthe, J., Kobzev, V., Styashkin, V., Achache, J., Slavin, J., Luhmann, J.G., Russell, C.T., 1989. Magnetic fields near Mars: first results. Nature 341 (6243), 604–607

35. Rosenbauer, H., Shutte, N., Apathy, I., Galeev, A., Gringauz, K., Grüwaldt, H., Hemmerich, P., Jockers, K., Kiraly, P., Kotova, G., Livi, S., Marsch, E., Richter, A., Riedler, W., Remizov, T., Schwenn, R., Schwingenschuh, K., Steller, M., Szegö, K., Verigin, M., Writte, M., 1989. Ions of martian origin and plasma sheet in the martian magnetosphere: initial results of the TAUS experiment. Nature 341 (6243), 612–614.
36. Selivanov, A.S., Naraeva, M.K., Panfilov, A.S., Gektin, Y.M., Kharlamov, V.D., Romanov, A.V., Fomin, D.A., Miroshnichenko, Y.Y., 1989. Thermal imaging of the surface of Mars. Nature 341 (6243), 593–595.
37. Shutte, N.M., Kiraly, P., Cravens, T.E., Dyachkov, A.V., Gombosi, T.I., Gringauz, K.I., Nagy, A.F., Sharp, W.E., Sheronova, S.M., Szegö, K., Szermerey, T., Szucs, I.T., Tatrallyay, M., Toth, A., Verigin, M., 1989. Observation of electron and ion fluxes in the vicinity of Mars with the HARP spectrometer. Nature 341 (6243), 614–616.
38. Simmons, K.E., Hendrix, A.R., 1999. Ultraviolet observations of Phobos, Deimos, Gaspra and Ida from Mariner 9 and Galileo. Bull. Astron. Soc. 31 (4), 1089–1090 (#11.05).
39. Sloan, R.K., 1968. Mariner Mars 1964 Project Report: Scientific Experiments. Jet Propulsion Laboratory, California Institute of Technology, TR 32-883, Pasadena, California
40. Steinbacher, R.H., Kliore, A., Lorell, J., Hipsher, H., Barth, C.A., Masursky, H., Münch, G., Pearl, J., Smith, B., 1972. Mariner 9 Scientific Experiments – Preliminary Results. Science 175 (4019), 293.
41. Steinbacher, R.H., Haynes, N.R., 1973. Mariner 9 mission profile and project history. Icarus 18 (1), 64–74.
42. Surkov, Y.u.A., Barsukov, V.L., Moskaleva, L.P., Kharyukova, V.P., Zaitseva, S.Y.E., Smirnov, G.G., Manvelyan, O.S., 1989. Determination of the elemental composition of martian rocks from Phobos 2. Nature 341 (6243), 595–598.
43. Thomas, N., Britt, D.T., Herkenhoff, K.E., Murchie, S.L., Semenov, B., Keller, H.U., Smith, P.H., 1999. Observations of Phobos, Deimos, and bright stars with the Imager for Mars Pathfinder. J. Geophys. Res. 104 (E4), 9055–9068, http://dx.doi. org/10.1029/98JE02555.
44. Thomas, N., Stelter, R., Ivanov, A., Bridges, N.T., Herkenhoff, K.E., McEwen, A.S., 2010. Spectral heterogeneity on Phobos and Deimos: HiRISE observations and comparisons to Mars Pathfinder results. Planet. Space Sci. 59 (13), 1281–1292, http://dx.doi.org/10.1016/j.pss.2010.04.018.
45. d0 Uston, C., Atteia, J.L., Barat, C., Chernenko, A., Dolidze, V., Dyatchkov, A., Jourdain, E., Khariukova, V., Khavenson, N., Kozlenkov, A., Kucherova, R., Mitrofanov, I., Moskaleva, L., Niel, M., Pozanenko, A., Scheglov, P., Surkov,

Y., Vilchinskaya, A., 1989. Observation of the y-ray emission from the martian surface by the APEX experiment. Nature 341 (6243), 598–600.

46. Valnicek, B., Farnik, F., Linkin, O.B., Pissarenko, N.F., Garcia, H., 1991a. Solar X-ray observation during the flight of Phobos 2 with the RF-15 instrument. Planet. Space Sci. 39 (1/2), 1–7.
47. Valnicek, B., Peresty, R., Zhitnik, I., 1991b. Results of the solar X-ray image study onboard the Phobos 1 space probe (the TEREK experiment). Planet. Space Sci. 39 (1/2), 7–13.
48. Witasse, O., Duxbury, T., Chicarro, A., Altobelli, N., Andert, T., Aronica, A., Barabash, S., Bertaux, J.L., Bibring, J.P., Cardesin-Moinelo, A., Cichetti, A., Companys, V., Dehant, V., Denis, M., Formisano, V., Futaana, Y., Giuranna, M., Gonde, B., Heather, D., Hoffmann, H., Holmström, M., Manaud, N., Martin, P., Matz, K.D., Montmessin, F., Morley, T., Mueller, M., Neukum, G., Oberst, J., Orosei, R., Pätzold, M., Picardi, G., Pischel, R., Plaut, J.J., Reberac, A., Pardo Voss, P., Roatsch, T., Rosenblatt, P., Remus, S., Schmedemann, N., Willner, K., Zegers, T., 2013. Mars Express investigations of Phobos and Deimos. Planet. Space Sci , http://dx.doi.org/10.1016/j.pss.2013.08.002i

CITATION

Thomas C. Duxbury, Alexander V. Zakharov, Harald Hoffmann, Edward A. Guinness, Spacecraft exploration of Phobos and Deimos, Planetary and Space Science, Volume 102, 1 November 2014, Pages 9-17, ISSN 0032-0633, http://dx.doi.org/10.1016/j.pss.2013.12.008.

Chapter 3

Rescheduling of Observing Spacecraft Using Fuzzy Neural Network and Ant Colony Algorithm

Li Yuqing, Wang Rixin, Xu Minqiang

Deep Space Exploration Research Center, Harbin Institute of Technology, Harbin 150080, China

ABSTRACT

This paper aims at rescheduling of observing spacecraft imaging plans under uncertainties. Firstly, uncertainties in spacecraft observation scheduling are analyzed. Then, considering the uncertainties with fuzzy features, this paper proposes a fuzzy neural network and a hybrid rescheduling policy to deal with them. It then establishes a mathematical model and manages to solve the rescheduling problem by proposing an ant colony algorithm, which introduces an adaptive control mechanism and takes advantage of the information in an existing schedule. Finally, the above method is applied to solve the rescheduling problem of a certain type of earth-observing satellite. The computation of the example shows that the approach is feasible and effective in dealing with uncertainties in spacecraft observation scheduling. The approach designed here can be useful in solving the problem that the original schedule is contaminated by disturbances.

INTRODUCTION

A growing fleet of observing spacecraft uses a variety of sensing technologies for scientific, mapping, defense, and commercial activities. Hundreds of observing tasks are required to be taken every day. As the number of spacecraft and the number of observation requests grow larger and larger, scheduling these tasks is more and more difficult. Foremost is the fact that the observation scheduling is a NP-complete problem. Furthermore, there are various uncertainties in spacecraft management and operation.

Automatic planning and scheduling for spacecraft observation is the key to promoting the efficiency of spacecraft management and reducing its operation costs. Hence, it has aroused the interests of scholars in fields such as aeronautics, computer science, and operation research. Potter et al.[1] studied the scheduling problem of a linear finite deterministic model using a backtracking method, and introduced its application in the Landsat 7 mission. Lin et al.[2] and Liao et al.[3] designed the daily imaging scheduling system of ROCSAT-II using methods of tabu search, random integer programming, and Lagrange relaxation. Lemaitre et al.4 and 5 studied the management of satellite resources and the observation scheduling of agile satellites. They made a further step to compare the performances of greedy algorithm, dynamic programming algorithm, and constraint programming algorithm. Wolfe et al.6, 7, 8, 9, 10 and 11 studied the application of genetic algorithm to the present problem. Globu et al.12 and 13 compared the performances of genetic algorithm, simulated annealing, and squeaky wheel optimization, and pointed out that stimulated annealing was the best technique in their study. In recent years, particle swarm optimization14 and 15 and heuristic search algorithm16, 17, 18 and 19 were applied into this domain. Besides, many other scholars also studied the problem of multi-satellite management from different perspectives.19, 20, 21, 22, 23 and 24

However, existing research on automatic planning and scheduling for spacecraft observation mainly focuses on how to establish an initial schedule in a deterministic environment while lacks considerations for the actual operation of a satellite which is in a dynamic environment full of uncertainties such as spacecraft fault and changes in observation demands.23 and 24 All of these uncertainties will affect the original schedule and cause rescheduling.

The purpose of this paper is to solve the rescheduling problem of spacecraft observation under uncertainties. What types of uncertainty are there in spacecraft observation? How do they affect the existing schedule? And how should we deal with these uncertainties? To answer these questions, the remainder of the paper is organized as follows. The next section makes a systematic analysis of uncertainties in spacecraft observation scheduling and points out that many of them have fuzzy features. Section 3 then proposes a fuzzy neural network to deal with these fuzzy uncertainties and a rescheduling policy to respond all kinds of uncertainties. Section 4 establishes a mathematical model and manages to solve this problem by using an ant colony algorithm. Finally, the above method is applied to solve the rescheduling problem of a certain type of earth-observing satellite in Section 5.

UNCERTAINTIES IN SPACECRAFT OBSERVATION

When operating a spacecraft, there are a variety of uncertainties which demand modifications of the original schedule. In order to make efficient rescheduling policies, it is necessary to analyze the uncertainties thoroughly.

There are many ways to categorize the uncertainties. Based on the actual operating conditions, the present paper divides the uncertainties into four types according to their origins: inherent uncertainties, uncertainties aroused by changes in internal and external environments, and discrete uncertainties, as listed in Table 1.

Table 1. Uncertainties in spacecraft observation.

Origins		Disturbance	Degree
Inherent uncertainties	Deviation of process time	I-Type gradual disturbance	Medium
	Orbit prediction error	I-Type gradual disturbance	Medium
Uncertainties aroused by changes in internal environments	Satellite fault	Abrupt disturbance	Serious
	Satellite recovery	Abrupt disturbance	Serious
	Breakdown of some payload	Abrupt disturbance	Serious
	Recovery of some payload	Abrupt disturbance	Serious
	Partial fault of some payload	II-Type gradual disturbance	Fuzzy
	Partial recovery of some payload	II-Type gradual disturbance	Fuzzy
	Breakdown of some resource	Abrupt disturbance	Serious
	Recovery of some resource	Abrupt disturbance	Serious
	Partial fault of some resource	II-Type gradual disturbance	Fuzzy
	Partial recovery of some resource	II-Type gradual disturbance	Fuzzy
Uncertainties aroused by changes in external environments	Observation condition is not optimum	II-Type gradual disturbance	Fuzzy
	Arrival of new tasks	II-Type gradual disturbance	Fuzzy
	Task cancelation	II-Type gradual disturbance	Fuzzy
	Priority change of existing tasks	II-Type gradual disturbance	Fuzzy
	Precision change of existing tasks	II-Type gradual disturbance	Fuzzy
	Payload change of existing tasks	II-Type gradual disturbance	Fuzzy
Discrete uncertainties	Demand of attitude and orbit control	Abrupt disturbance	Serious
	Other demands of control	Abrupt disturbance	Serious

In order to deal with various uncertainties in proper ways, it is not enough to categorize them only according to their origins. There are 20 uncertainties listed in Table 1. According to their disturbances to the system, the uncertainties are divided into three categories: abrupt disturbance, I-Type gradual disturbance, and II-Type gradual disturbance. Abrupt disturbance refers to the uncertainties with a great amplitude of changes. These uncertainties impose serious effects on the system and in most cases need a new schedule. I-Type gradual disturbance refers to the uncertainties that change frequently but mildly, which can be dealt with by periodic rescheduling. II-Type gradual disturbance bears a similarity with abrupt disturbance in their origins, but the disturbance degree of the former is not as serious. Due to its fuzzy feature, the degree of II-Type gradual disturbance needs to be evaluated according to specific conditions.

It is easier to process abrupt disturbance and I-Type gradual disturbance. They can be dealt with by using periodic and event-driven rescheduling policies. Whereas it is complicated to process II-Type gradual disturbance due to the following reasons:

1) It is difficult to judge the responding time. If these uncertainties are not responded in time, the operating efficiency will be low and this will result in insufficient use of spacecraft resources. If these uncertainties are always responded in time, the computing load of the system will be large and the stability of the system will be affected.
2) It is difficult to judge its degree of disturbance to the system. Firstly, there are many parameters which affect the calculation of the degree of disturbance. Secondly, it is hard to precisely decide the values of many parameters because of their fuzzy features.

Due to the preceding two reasons, it is difficult to process II-Type gradual disturbance using an existing rescheduling method. The key to the problem is how to judge its degree of disturbance to the system and how to respond properly.

EVALUATION OF DISTURBANCE DEGREES OF UNCERTAINTIES

Fuzzy neural network, due to its intelligence and learning capability, has been widely used in decision-making systems. This paper uses this method to evaluate the degree of II-Type gradual disturbance. The method consists of two parts: fuzzy processing of parameters and neural network. The degree of membership of parameters can be obtained by applying fuzzy processing to original data and function as the input of the fuzzy neural network. The output of the network is the result of decision, i.e., the degree of disturbance to the system. Evaluation rules can be learned by training of the neural network and reflected by the weights of the neural network. Quantization of the parameters that can influence decision-making can be realized by the computation of the degree of membership on the input layer.

Fuzzy Processing of Parameters

Let δ be the degree of II-Type gradual disturbance to the system. Then parameters influencing the value of δ fall into five categories: types of uncertainties, intensity of uncertainties, urgency of uncertainties, expected earning of responding to uncertainties, and the number of accumulated uncertainties. See Table 2.

Table 2. Parameters of II-Type gradual uncertainty.

Parameter	Domain	Value/fuzzy subsets
Types of uncertainties	C	{e, r, c, tn, tp, tr, te}
Intensity of uncertainties	I	$\{I_s, I_w, I_m, I_a, I_c\}$
Urgency of uncertainties	U	$\{U_s, U_w, U_m, U_a, U_c\}$
Expected earning	W	$\{W_w, W_m, W_a\}$
Accumulated uncertainties	A	$\{A_w, A_m, A_a\}$

Domain I, U, W, and A all have a fuzzy feature except C. The fuzzy subsets can be defined as follows:

$T(I) = \{I_s, I_w, I_m, I_a, I_c\}$, indicating that the intensity of uncertainties is slight, weak, moderate, acute, critical.

$T(U) = \{U_s, U_w, U_m, U_a, U_c\}$, indicating that the urgency of uncertainties is slight, weak, moderate, acute, critical.

$T(W) = \{W_w, W_m, W_a\}$, indicating that the expected earning is low, moderate, high.

$T(A) = \{A_w, A_m, A_a\}$, indicating that the number of accumulated uncertainties is small, moderate, large.

The detail of e,r,c,tn,tr,te is given in Table 3. Taking the intensity of uncertainties as an example, the evaluation of this parameter depends on the number and priority of affected tasks. Then it can be defined as:

$$\mu_I = \frac{1}{2}\left(\frac{n_{\text{affected}}}{N} + \frac{\text{pri}_{\text{affected}}}{\text{PRI}}\right) \quad (1)$$

where n_{affected} is the number of affected tasks, N is the total number of tasks, $\text{pri}_{\text{affected}}$is the sum of priority of affected tasks, and PRI is the sum of priority of all the tasks. The degree of membership of I_s, I_w, I_m, I_a, and I_c can be obtained on a 0–100% scale of μ_I as shown in Fig. 1.

Table 3. Categories of II-Type gradual uncertainty.

Value of uncertainty type	Detail
e	Fault of some payload
	Partial recovery of some payload
r	Fault of some resource
	Partial recovery of some resource
c	Observation condition is not optimum
tn	Arrival of new tasks
	Task cancelation
tp	Priority change of existing tasks
tr	Precision change of existing tasks
te	Payload change of existing tasks

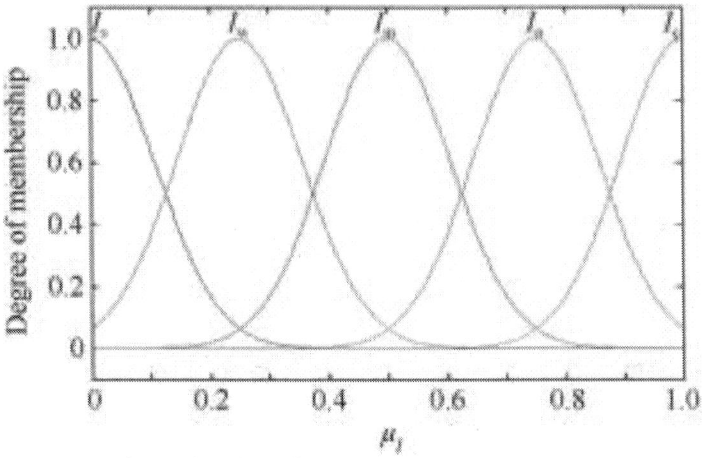

Figure. 1. Degree of membership of intensity.

Let ΔT_{min} be the minimum rescheduling interval, and according to the number of affected tasks during ΔT_{min}, the system can evaluate the urgency of uncertainties *T(U)*. Similarly, the degree of membership of elements in *T(W)* and *T(A)* can be obtained according to the expected earning and the number of accumulated tasks, respectively.

Neural Network

As mentioned above, the neural network is the second part. Both BP neural network and radial basis function (RBF) neural network are commonly used in decision-making systems. However, RBF neural network uses Gauss function as the activation function and belongs to local approximation neural network. Therefore, the convergence rate of RBF neural network is faster than that of BP neural network. Thus, when evaluating the degree of II-Type gradual disturbance to the system, RBF neural network is selected. See Fig. 2.

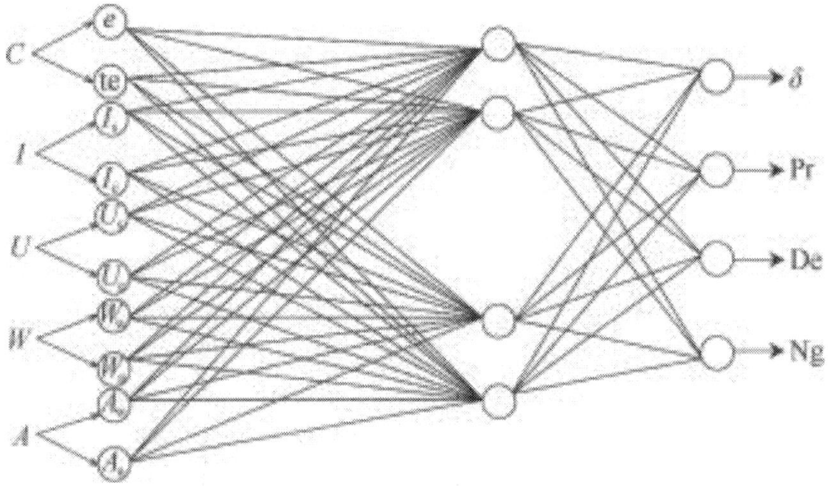

Figure. 2. Diagram of radial basis function neural network.

Firstly, the input and output of the RBF neural network must be determined. Let the 5 parameters mentioned in Table 2 be the input of the RBF neural network. In these 5 parameters, the parameter of uncertainty type does not have a fuzzy feature, but it has a concurrency feature. For the RBF neural network input, this parameter is defined as a finite set with 7 elements. See Table 3.

Rescheduling demands caused by various uncertainties can be classified into 5 kinds according to their origins and responding policies, 3 of which (Pr, De, and Ng) are caused by II-Type gradual disturbance. See Table 4. Let Pr, De, Ng, and δ be the output of the RBF neural network.

Table 4. Categories of rescheduling demands.

Kinds of rescheduling demands	Causes
Periodic rescheduling *P*	Finished execution of existing schedule
Abrupt rescheduling Ab	Abrupt disturbance
Prompt rescheduling Pr	II-Type gradual disturbance
Delayed rescheduling De	II-Type gradual disturbance
Negligible rescheduling Ng	II-Type gradual disturbance

As shown in Fig. 2, the RBF neural network consists of an input layer, a hidden layer, and an output layer. The activation function of node uses Gauss function. See Eq. (2).

$$R_j(\boldsymbol{x}) = \exp\left(-\frac{\|\boldsymbol{x}-\boldsymbol{t}_j\|^2}{2\sigma_j^2}\right) \quad (j = 1, 2, \ldots, s) \tag{7}$$

where $\boldsymbol{x}$ is an m-dimensional input vector; $\boldsymbol{t}_j$ is the mean vector of node j on the hidden layer; σ_j is the deviation of node j on the hidden layer; s is the number of hidden layer nodes. The input layer implements nonlinear mapping from $\boldsymbol{x} \rightarrow R_j(\boldsymbol{x})$; the output layer implements linear mapping from $R_j(\boldsymbol{x}) \rightarrow y_k(\boldsymbol{x})$ and $y_k(\boldsymbol{x})$ is formulated as:

$$y_k(\boldsymbol{x}) = \sum_{j=1}^{s} \omega_{kj} R_j(\boldsymbol{x}) \quad (k = 1, 2, \ldots, l) \tag{3}$$

where l is the number of output layer nodes and ω_{kj} is the weights.

The learning process of the RBF neural network is a process that minimizes the mean square error via adjusting weights and thresholds of the network. The mean square error is formulated as Eq. (4); the parameters are adjusted as Eqs. (5) and (6):

$$F(\boldsymbol{x}) = E[e(\boldsymbol{x})] = (t(\boldsymbol{x}) - y(\boldsymbol{x}))^T (t(\boldsymbol{x}) - y(\boldsymbol{x})) = e^T(\boldsymbol{x}) e(\boldsymbol{x}) \tag{4}$$

$$w_{ij}(n+1) = w_{ij}(n) - \alpha \frac{\partial F}{\partial w_{ij}} \quad (i = 1, 2, \ldots, s;\ j = 1, 2, \ldots m) \tag{5}$$

$$b_i(k+1) = b_i(k) - \alpha \frac{\partial F}{\partial b_i} \quad (i = 1, 2, \ldots s) \tag{6}$$

where $t(\boldsymbol{x})$ is the expected output of the network, $y(\boldsymbol{x})$ is the actual output of the network, w_{ij} is the weight, b_i is the threshold, α is the learning rate.

Rescheduling Policy

In order to make proper responses to various rescheduling demands, a hybrid rescheduling policy based on the minimum interval is proposed.

Different responding ways are adopted according to different kinds of rescheduling demands.

When the trigger condition of periodic and abrupt rescheduling demand is met, the system will make an immediate response.

When the rescheduling demand is caused by II-Type gradual disturbance, the responding ways depend on the output of the neural network.

If the output is Pr, it needs to be responded immediately because of its serious disturbance to the system.

If the output is De, the responding ways depend on Δt and ΔT_{min}, where Δt is the interval between this rescheduling demand and the last rescheduling responding and ΔT_{min} is the minimum rescheduling interval: if $\Delta t < \Delta T_{min}$, then the response is delayed until $\Delta T_{min}$is satisfied; otherwise, it will be responded immediately.

If the output is Ng, the rescheduling demand will be ignored because of its slight disturbance to the system.

MATHEMATICAL STATEMENT AND ANT COLONY ALGORITHM

In its observations, a spacecraft often carries several kinds of payload to perform different observing tasks. These tasks must satisfy some constraints, such as power availability, limited imaging segments per orbit, and revisit limitations. During the implementation of the schedules, there is a possibility that the existing tasks may be modified or canceled, or that a new task may arrive. Besides, there may arise faults in some payloads. Since there are so many uncertainties in the actual operation of a spacecraft, it is necessary to study spacecraft observation rescheduling under uncertainties.

In this section, an ant colony algorithm is studied, and the purpose is to find a way to establish a proper mathematical model which could be resolved by any colony algorithm.

Mathematical Statement of Spacecraft Observation Rescheduling

Let SAT_{power} be the amount of available power of the spacecraft and SAT_{ssr} the total memory size on board. Under normal circumstances, they are constants that do not change with time. When faults occur, they might be influenced and thus change. Let $p_1, p_2, \ldots, p_H$ be H payloads; $o_1, o_2, \ldots, o_N$ be N observing requests which are the elements of an observing request set O during interval $[t_{start}, t_{end}]$. Each element o_i in O carries the following attributes:

o_i_C: selected or not.
o_i_A: longitude and latitude coordinates.
o_i_P: payload.
o_i_D: observation duration.
o_i_V: weight.
o_i_W: set of visible intervals.
o_i_S: starting time of observation.
o_i_E: ending time of observation.
o_i_POWER: power needed for observation.
o_i_SSR: memory size needed for observation.
o_i_Pri: priority.

The task of communication with ground stations is also defined as an observation request, but the only difference lies between it and other observation requests is thato_i_SSR is negative.

Because the purpose of observing spacecraft is to maximize the amount of useful return data, the quality criterion is formulated as:

$$\mathrm{Max}\left\{\sum_{i=1}^{N} o_i\_C \times o_i\_V\right\} \tag{7}$$

where $o_i\_C = \{0,1\}$. if selected, $o_i\_C = 1$, if not selected, $o_i\_C = 0$, o_i_V is the observation weight.
Constraints:

(1) Resource constraints

$$\sum_{o_i \in O} o_i\_C \times o_i\_\mathrm{SSR} \leq \mathrm{SAT}_{\mathrm{ssr}} \tag{8}$$

$$\forall t \in [t_{\text{start}}, t_{\text{end}}],\ \exists \sum_{o_i\_S<t<o_i\_E} o_i\_C \times o_i\_\text{POWER} \leq \text{SAT}_{\text{POWER}} \quad (9)$$

(1) Temporal constraints

$$\begin{aligned}&\forall o_i \in O, \text{if } o_i\_C = 1, \text{ then } \exists o_i\_w_j \in o_i\_W,\\ &\text{subject to } o_i\_S \in o_i\_w_j \text{ and } o_i\_E \in o_i\_w_j\end{aligned} \quad (10)$$

$$\forall o_i \in O, \text{if } o_i\_C = 1 \text{ then } o_i\_S \in [t_{\text{start}}, t_{\text{end}}] \text{ and } o_i\_E \in [t_{\text{start}}, t_{\text{end}}] \quad (11)$$

$$\forall o_i, o_j \in O, o_i\_S < o_j\_S \text{ if } o_i\_P = o_j\_P \text{ then } o_i\_E < o_j\_S \quad (12)$$

where Eq. (8) indicates that the memory size needed for all selected requests must be smaller than the total memory size on board; Eq. (9) indicates that the available power must be able to support task implementation at any time; Eq. (10) indicates that if an observing request is selected, then the target must be observed during interval $o_i_w_j$which is the j th element of set o_i_W. Eq. (11) indicates that all tasks selected must be finished during interval [t_{start}, t_{end}]. Eq. (12) indicates that each payload implements an observing activity once a time, i.e., the next observing activity begins after its forgoing activity gets finished.

Rescheduling based on ant Colony Algorithm

Basic ant Colony Observation Scheduling Optimization

According to the mathematical model, spacecraft rescheduling is a 0–1 integer planning problem which is an NP complete problem. The polynomial time algorithm hasn't been found yet. Swarm Intelligence algorithms, particularly the ant colony algorithm, have been used to optimize a wide variety of tasks.[25] Therefore, this paper attempts to solve the rescheduling problem of observing spacecraft using this algorithm.

Firstly, to avoid solving the problem in an infinite domain, observing time must be discretized. As shown in Fig. 3, o_i_W is the set of all visible intervals of observing requesto_i. In actual operations, observation can be performed at any time during an interval, but this will result in searching in an infinite domain. To avoid this, defining the time required by one observation as a unit, o_i_W could be discretized as shown in Fig. 3. That is, the observation must be implemented in one or several units.

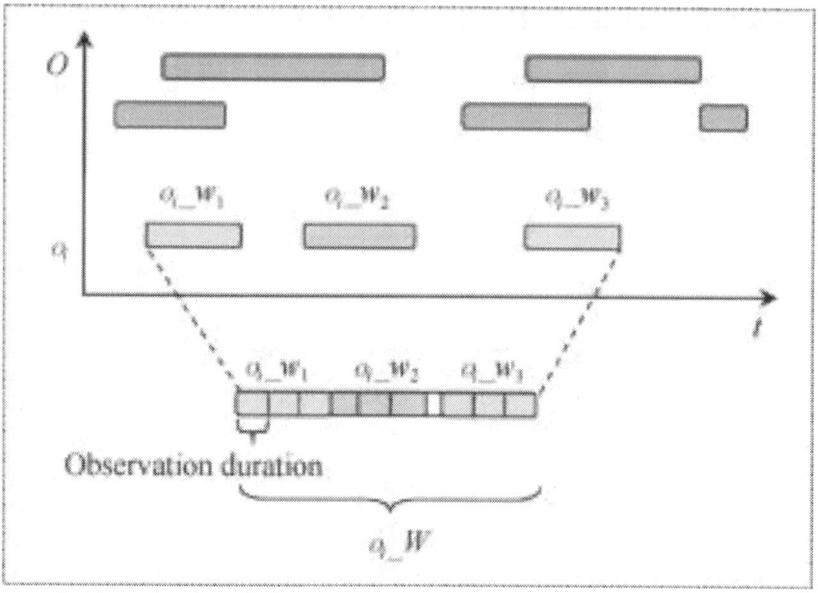

Figure. 3. Discretization of observing time.

LET o_i in set O be the node processed by the ant colony algorithm. Let $b_i(t)$ (i = 1,2, ..., N) be the ant number in node i at time t . Let $m = \sum_{i=1}^{N} b_i(t)$ be the total number of ants, and each ant has features as follows:

1) It expands the next node by the probability which is computed according to the pheromones. (Let $\tau_i(t)$ (i = 1,2, ..., N) be the pheromones in node i at time t.)
2) Assume each node can be visited only once. This rule is controlled by tabu_k, where$k$ = 1,2, ..., m. (Let tabu_k be the taboo list of ant k.)
3) When travel is finished, each ant will leave pheromones at every node that it has visited.

The pheromones of each node is the same at initial time, i.e., $\forall o_i \in O, \exists \tau_i(0) = \text{Cont}$, where Cont is a constant and *i* = 1,2, ..., *N*. Ant *k* (*k* = 1,2, ..., *m*) expands the next node by $p_j^k(t)$, which is the probability of selecting node *j* at time *t*:

$$p_j^k(t) = \begin{cases} \frac{\tau_j(t)}{\sum_{s \in \text{allowed}_k} \tau_s(t)} & \text{If } j \in \text{allowed}_k \\ 0 & \text{Others} \end{cases} \quad (13)$$

where allowed$_k$ = *O*–tabu$_k$, which is the nodes set of ant *k* that can be selected at the next step. A new node will be selected if all the constraints can be satisfied. Otherwise, check another node. When a loop is finished after *N* steps, the pheromones of each node are updated as follows:

$$\tau_i(t+N) = \rho\tau_i(t) + \Delta\tau_i \quad (14)$$

$$\Delta\tau_i = \sum_{k=1}^{m} \Delta\tau_i^k \quad (15)$$

$$\Delta\tau_i^k = \begin{cases} \frac{Q}{L_k}, & \text{If node } i \text{ is selected by ant } k \\ 0, & \text{Others} \end{cases} \quad (16)$$

where $\Delta\tau_i^k$ is the pheromones that are left at node *i* by ant *k* in the current loop, $\Delta\tau_i$ is the increased pheromones in the current loop, ρ is the pheromone evaporation rate,*Q*is the pheromone intensity, L_k is the fitness of ant *k*.

The ant colony algorithm is executed as follows.

Ant colony observation Scheduling Optimization (ACOSO):

Step 1. Let nc = 0. (nc is the number of iteration times.) Initialize τ_i and $\Delta\tau_i$, and place *m*ants at *N* nodes.

Step 2. Each ant expands the next node according to the probability $p_j^k(t)$, and loop this step until all nodes are visited.

Step 3. Calculate L_k of ant k (k = 1,2, ..., m), and record the best solution.

Step 4. Update the pheromones of each node.

Step 5. nc = nc + 1.

Step 6. If nc < NC (NC is the maximum iterations) and the solution is not degenerate, then go to step 2.

Step 7. Output the best solution.

Adaptive Ant Colony Observation Scheduling Optimization

In ACOSO, the basic ant colony algorithm is adopted. In order to improve the performance of the algorithm, this paper introduces the adaptive mechanism into the basic ant colony. The new algorithm is called adaptive ant colony observation scheduling optimization (AACOSO). Base on the existing research,[25] there are two important parameters which are hard to set, the pheromone intensity Q and the evaporation rate ρ.

The pheromone intensity Q is a constant that expresses the pheromone amount left by an ant, and the evaporation rate ρ is a constant that expresses the pheromone volatilization speed. Both of them can affect the convergence and the solution accuracy. That is, a larger Q or ρ can make a faster convergence but a higher risk on a local optimal solution. On the contrary, a smaller Q or ρ can find a better solution beyond the local optimal but a poor convergence follows. In order to make use of this feature, this paper sets Q as follows:

$$Q(t) = \begin{cases} Q_{\max} & t = 0 \\ \min\{Q(t-1)[1+0.1q(t)], Q_{\max}\} & \text{if find a better solution} \\ \max\{Q(t-1) - [Q(t-1) - Q_{\min}][0.1 \times q(t)], Q_{\min}\} & \text{if stagnate } N_Q \text{ times} \\ Q(t-1) & \text{Otherwise} \end{cases} \tag{17}$$

where $Q(t) \in [Q_{min}, Q_{max}]$, and $q(t)$ is adapted as in Eq. (18).

$$q(t)=\begin{cases} q_{\min} & t=0 \\ \max\{q(t-1)-1, q_{\min}\} & \text{If find a better solution} \\ \min\{q(t-1)+1, q_{\max}\} & \text{If stagnate } N_q \text{ times} \\ q(t-1) & \text{Otherwise} \end{cases} \tag{18}$$

where $q(t) \in [q_{\min}, q_{\max}]$.

It can be drawn from Eqs. (17) and (18) that Q is $Q_{\max}$ in the beginning. This is good for a quickly search and convergence. When the algorithm is stuck in a local optimal solution, AACOSO will decrease the value of Q so that the pheromone increasing rate and positive feedback are reduced. This is good for finding a better solution.

Similarly, the evaporation rate ρ is set as follows:

$$\rho(t)=\begin{cases} \rho_{\max} & t=0 \\ \min\{\rho(t-1)\times[1+0.1\times p(t)], \rho_{\max}\} & \text{if find a better solution} \\ \max\{\rho(t-1)-[\rho(t-1)-\rho_{\min}]\times[0.1\times p(t)], \rho_{\min}\} & \text{if stagnate } N_p \text{ times} \\ \rho(t-1) & \text{Otherwise} \end{cases} \tag{19}$$

where $\rho(t) \in [\rho_{\min}, \rho_{\max}]$, and $p(t)$ is adapted as in Eq. (20).

$$p(t)=\begin{cases} p_{\min} & t=0 \\ \max\{p(t-1)-1, p_{\min}\} & \text{if find a better solution} \\ \min\{p(t-1)+1, p_{\max}\} & \text{if stagnate } N_p \text{ times} \\ p(t-1) & \text{Otherwise} \end{cases} \tag{20}$$

where $p(t) \in [p_{\min}, p_{\max}]$.

Additionally, when unexpected events occur, in most cases, there are many tasks that won't be affected by the disturbance. Taking Fig. 4 for example, assume that there is an uncertainty called disturbance A occurred at t_0, and only tasks *a*, *b*, *c* and *d* are affected while others are not. It may be more effective if taking this information into account when rescheduling.

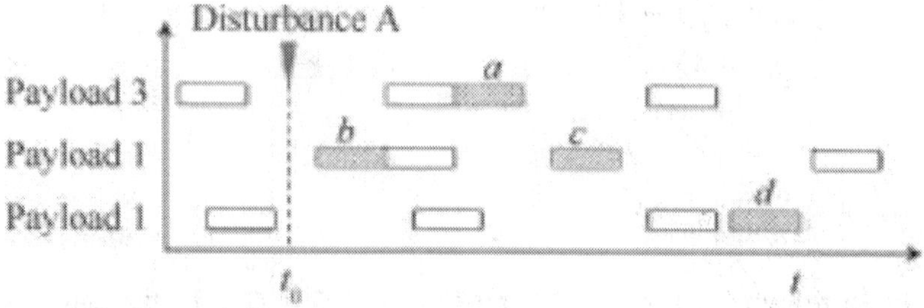

Figure. 4. Illustration of partially-affected schedule.

It can be calculated that which task is affected based on the temporal constrains and the resource constrains. In order to make use of this information, AACOSO only updates the pheromones of affected tasks with setting them to a constant and preserves those of unaffected ones. The detail of the rescheduling algorithm is as follows.

Adaptive Ant Colony Observation Scheduling Optimization (AACOSO):

Step 1. Calculate and determine the tasks which are affected. Update the observing request set *O*.

Step 2. Initialize the pheromones of the new nodes, and preserve those of the old ones.

Step 3. Let nc = 0, and place *m* ants at *N* nodes.

Step 4. Each ant expands the next node according to the probability $p_j^k(t)$, and loop this step until all nodes are visited.

Step 5. Compute Q_k of ant *k* (*k* = 1,2, …, *m*), and record the best solution.

Step 6. Update the pheromones of each node.

Step 7. nc = nc + 1.

Step 8. If nc < NC and the solution does not degenerate, then go to Step 4.

Step 9. Output the best solution.

NUMERICAL EXAMPLE

In this section, to demonstrate the validity of the proposed rescheduling policy and the new ant colony algorithm, the rescheduling problem of a certain type of earth-observing satellite is studied. Assume that the satellite carries 5 kinds of payloads and a solid state recorder with a 378 Gbits capacity. The details of the orbit and satellite parameters are shown in Table 5 and Table 6.

Table 5. Orbit parameters of the simulation case.

Parameters	Value
a (km)	7193.5899
e	0.002344
i (°)	98.726
Ω (°)	77.105
ω (°)	107.085
ϑ (°)	184.645

Table 6. Satellite parameters of the simulation case.

Satellite parameter	Value Power (W)	Data rate (Mbits/s)
Payload 1	590	150
Payload 2	500	90
Payload 3	90	2
Payload 4	670	100
Payload 5	400	10
Capacity	378 Gbit	
Data rate	105 Mbits/s	
Attitude control	Three-axis stabilization	
Rotational speed	1 (°)/s	
Ground station	5	

The goal is to schedule the observing requests that should be implemented in the next 6 h. The attributes of these requests such as longitude and latitude coordinates, payload, value, and observation duration are generated randomly. Some of them are listed in Table 7.

Table 7. Observation requests.

Observing request	o_i_P	o_i_V	o_i_D	o_i_W	
				Start	End
1	1	77	50	1398	1500
				7185	7244
2	1	81	40	1623	1701
				17 721	17 821
3	1	85	50	9953	9981
				12 904	13 019
4	4	70	60	17 728	17 826
5	2	86	50	12 076	12 173
				14 850	14 949
6	2	90	50	12 085	12 167
				14 862	14 967
				17 641	17 702
7	5	92	40	6268	6359
				16 642	16 715
8	5	87	50	12 039	12 097
				17 777	17 881
9	3	60	50	1398	1500
				7185	7244
10	3	78	50	19 470	19 523
11	1	62	60	19 994	20 075
12	2	77	50	12 082	12 170
				17 837	17 942
13	3	83	40	6152	6192
				16 290	16 403
14	1	80	40	12 070	12 133
				17 811	17 913
15	3	71	50	12 124	12 209
				17 880	17 976
⋮	⋮	⋮	⋮	⋮	⋮
166	5	92	50	12 946	12 066
167	3	81	40	14 700	14 789
168	1	66	50	14 700	14 789

169	5	72	40	456	545
170	2	64	60	7415	7529
				18 444	18 549
171	1	79	50	15 012	15 098
172	4	75	60	1684	1771
173	2	75	50	1684	1771
174	4	67	40	17 391	17 498
175	4	68	50	6155	6196
				21 292	21 406
176	5	79	50	6155	6196
				21 292	21 406
177	3	69	60	21 081	21 178
178	2	88	40	12 076	12 142
				17 819	17 920
179	1	86	50	7868	7902
180	1	73	40	12 593	12 670

These observing requests are scheduled by using the ant colony algorithm. The initial schedule is shown in Fig. 5.

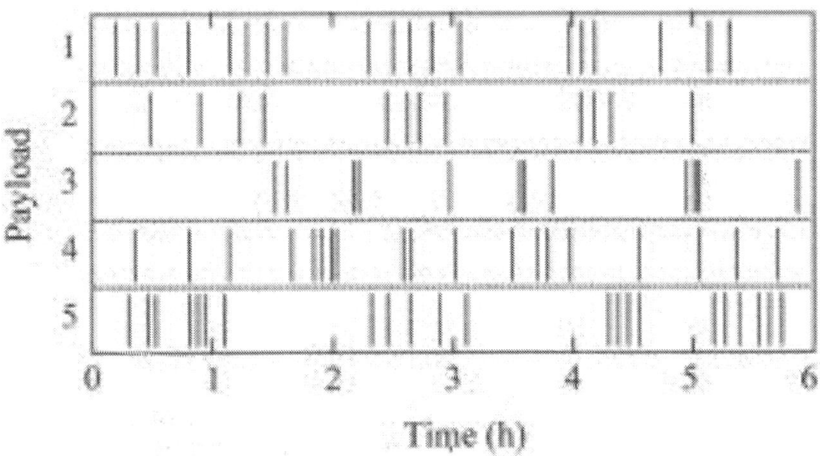

Figure. 5. Initial schedule.

When operating a spacecraft, there are a variety of uncertainties which demand modifications of the original schedule. Assume the rescheduling period is 2 h; let the minimum rescheduling interval ΔT_{min} be 30 min.

Assume there are 6 disturbances. Then, there will be 9 rescheduling requests as shown in Table 8.

Table 8. Calculated results.

Rescheduling request	Uncertainty	Occurring time	Kind	Output of the RBF neural network			
				δ	Pr	De	Ng
1	Fault of payload 1	0 h 52 min	Prompt rescheduling	0.87	0.9367	0.4981	0.1645
2	Cancelation of some tasks	1 h 23 min	Delayed rescheduling	0.77	0.3388	0.9667	0.1565
3	Arrival of rescheduling period	2 h	Periodic rescheduling	–	–	–	–
4	Fault of power unit 1	2 h 13 min	Negligible rescheduling	0.21	0.4361	0.3191	0.7946
5	Recovery of payload 1	3 h 14 min	Abrupt rescheduling	–	–	–	–
6	Arrival of rescheduling period	4 h	Periodic rescheduling	–	–	–	–
7	Arrival of new tasks	5 h 27 min	Delayed rescheduling	0.61	0.1542	0.8971	0.2271
8	Arrival of new tasks	5 h 35 min	Prompt rescheduling	0.61	0.8656	0.3412	0.1247
9	Arrival of rescheduling period	6 h	Periodic rescheduling	–	–	–	–

Using the proposed rescheduling policy, the system will reschedule 5 times. After that, the Gantt chart of the actual schedule is presented in Fig. 6. The Gantt chart shows that due to the fault of payload 1 at 0 h 52 min, the prompt rescheduling is triggered, and the relevant tasks between 0 h 52 min and 3 h 14 min are canceled, hence releasing some memory and power. Consequently, some tasks which gain observing opportunities are distributed to payloads 2, 3, and 4.

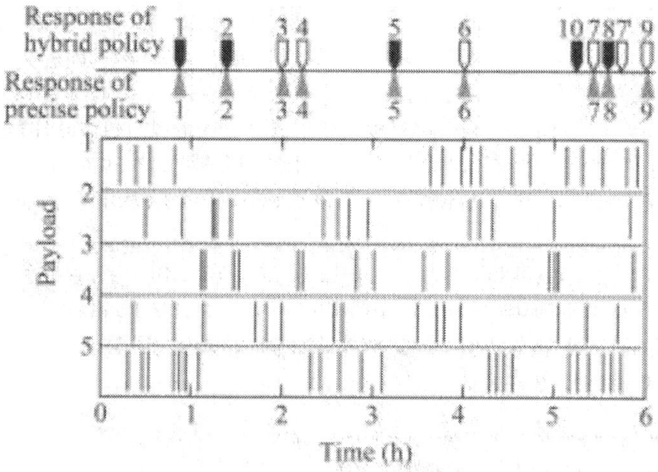

Figure. 6. Rescheduling results.

Disturbance 2 occurs at 1 h 23 min, and the delayed rescheduling is triggered. However, $\Delta t > \Delta T_{min}$, so the rescheduling is implemented immediately. Hence the tasks performed by payloads 3 and 4 are canceled, but there is no appropriate task that could be added, so there is no other change. Periodic rescheduling requests 3, 6, and 9 at 2 h, 4 h, and 6 h are canceled because of the rescheduling caused by disturbances 2, 5, and 8.

The influence on the system caused by disturbance 4 is slight because of the employment of redundancy unit, so the rescheduling request is canceled.

Because of the abrupt rescheduling caused by disturbance 5, some new tasks are distributed to payload 1, and some of the tasks performed by other payloads are adjusted correspondingly.

The periodic rescheduling is triggered at 5 h 14 min whose serial number is 10.

The delayed rescheduling request caused by disturbance 7 at 5 h 27 min is delayed to 7′, i.e., 5 h 44 min, because $\Delta t < \Delta T_{min}$.

The prompt rescheduling request caused by disturbance 8 at 5 h 35 min is implemented immediately due to its serious disruption on the system, and the effect caused by disturbance 7 is processed at the same time. Hence the rescheduling request 7′ is canceled.

For comparison, let the scheduling system respond every disturbance. This is called the precise rescheduling policy. Then the number of rescheduling times is 9. The result is the same as the above one. (See Fig. 6.) A conclusion can be drawn that the hybrid rescheduling policy proposed in this paper could save about 44.4% running time with the same performance compared with the precise rescheduling policy.

To validate the feasibility and validity of the ant colony algorithm, taking the rescheduling caused by disturbance 1 as an example, computation is performed using ACOSO and AACOSO separately and is run 10 times. The results are shown in Fig. 7.

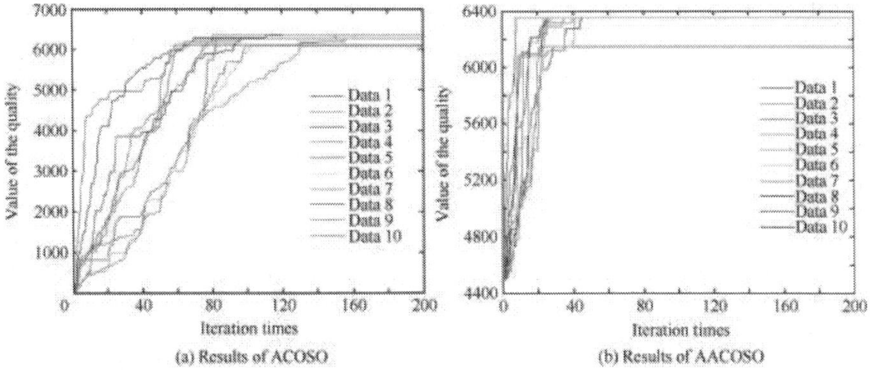

Figure. 7. Rescheduling caused by disturbance 1.

Fig. 7 shows that the probability of finding the optimum solution of AACOSO is 30% higher than that of ACOSO, and the speed of convergence of the former is obviously faster. See Table 9.

Fig 8 shows the evolutions of parameters Q and ρ. It can be seen from it that Q and ρ are both the maximum in the beginning. This is good for a quickly search and convergence. When the algorithm is stuck in a local optimal solution, AACOSO decreases the values of$Q$ and ρ to help find a better solution, and regains them after the better one is found. Because Q and ρ use similar rules to adjust their values, the evolution curves are similar as well.

Table 9. Results comparison.

Number	Value of the quality criterion	
	ACOSO	AACOSO
1	6348	6145
2	6265	6348
3	6093	6348
4	6112	6348
5	6348	6348
6	6348	6348
7	6348	6348
8	6112	6348
9	6348	6348
10	6348	6348
Frequency of finding the optimum solution	6	9
Speed of convergence	Slow	Fast

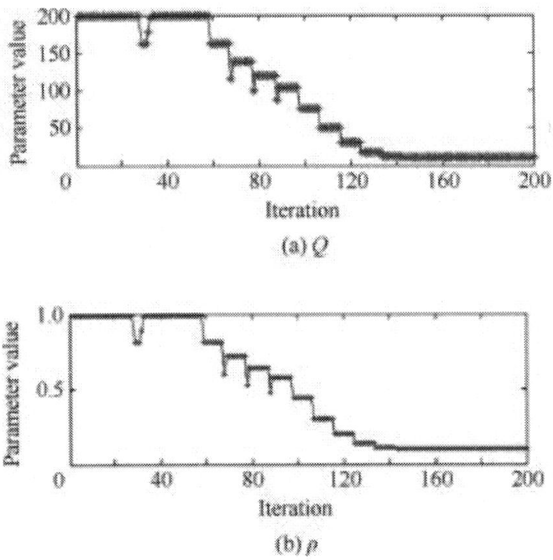

Figure. 8. Evolutions of the parameters Q and ρ.

In the above instance, the number of affected tasks at each rescheduling is different. In order to validate AACOSO more precisely, this paper uses different affected task numbers to test. Assume there are 1000 tasks need to be observed and there is one rescheduling. Set affected task numbers from 10 to 1000, and reschedule them by ACOSO and AACOSO separately. Run 20 times and take the average. The results are shown in Fig. 9.

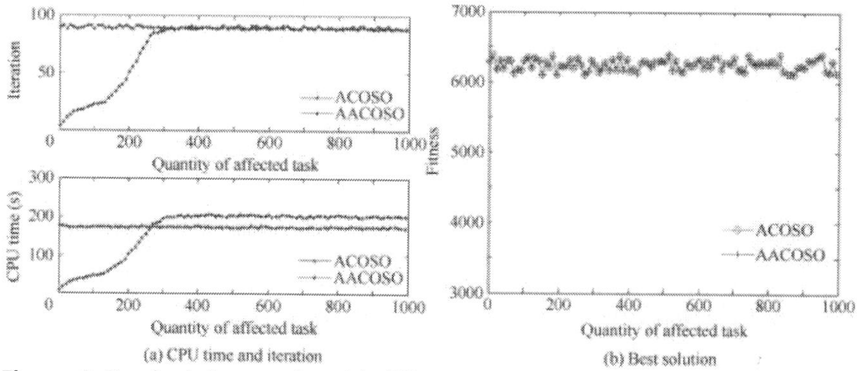

Figure. 9. Rescheduling results with different quantities of affected tasks.

It can be seen from Fig. 9 that if the number of affected tasks is less than 25% of the total, AACOSO is better than ACOSO, with less CPU time and iteration, and both of these can find the best solution. With the increasing of the affected tasks, the iteration is almost the same as that of ACOSO but the CPU time is higher. The reason is that AACOSO must calculate which task is affected and this is time costly when there are too many tasks. However, AACOSO can give a better solution more frequently (see Fig. 7), even with the higher CPU time.

CONCLUSIONS

In this paper, a method based on an ant colony algorithm and a fuzzy neural network is proposed to solve the rescheduling of spacecraft observation under uncertainties. Firstly, this paper makes a systematic analysis about uncertainties in spacecraft observation scheduling, and points out that many of them have fuzzy features. Then this paper proposes a fuzzy neural network to deal with these fuzzy uncertainties and a rescheduling policy to respond all kinds of uncertainties. Furthermore, this paper establishes a mathematical model and manages to solve this problem by an ant colony algorithm. The rescheduling problem of a certain type of earth-observing satellite is solved to validate the feasibility and validity of the proposed approach.

The fuzzy neural network can respond the fuzzy uncertainties appropriately. With the hybrid rescheduling policy, it can save about 44.4% running time and obtain the same solution as the precise rescheduling policy.

The ant colony algorithm is a new method to deal with the rescheduling problem of observing spacecraft. By introducing an adaptive control mechanism and taking advantage of the existing schedule, the probability of finding the optimum solution can be raised about 30%, and the speed of convergence is faster.

ACKNOWLEDGEMENTS

This work was supported by the National Natural Science Foundation of China (No.61203151), the National Basic Research Program of China (973 Program) (No.2012CB720003), the Postdoctoral Science Foundation of China (20100471044), the Fundamental Research Funds for the Central Universities of China (No.HIT.NSRIF.2013038), and the Key Laboratory Opening Funding of China (No.HIT.KLOF.2009071).

REFERENCES

1. Potter W, Gasch J, Bauer C. A photo album of Earth scheduling daily Landsat 7 mission daily activities. In: Proceedings of the international symposium space mission operations and ground data system, 1995 Jun 1–5; Tokyo, Japan; 1998. p. 1–7. http://track.sfo. jaxa.jp/spaceops98/paper98/ track2/ 2b010.pdf, http://track.sfo.jaxa. jp/spaceops98/paper98/track2/2b014.pdf.
2. Lin WC, Liao DY, Liu CY, Lee YY. Daily imaging scheduling of an earth observation satellite. IEEE Trans Syst Man Cybern A Syst Hum 2005;35(2):213–23.
3. Liao DY, Yang YT. Imaging order scheduling of an earth observation satellite. IEEE Trans Syst Man Cybern C Appl Rev 2007;37(5):794–802.
4. Lamaitre M, Verfaillie G, Bataille N. Sharing the use of a satellite: an overview of methods. In: Proceedings of SpaceOps, 1998. p. 1–7. http://track.sfo.jaxa.jp/spaceops98/paper98/track2/2b010.pdf, http://track.sfo.jaxa.jp/spaceops98/paper98/track2/2b014.pdf.
5. Lamaitre M, Verfaillie G, Jouhaud F, Lachiver JM, Bataille N. Selecting and scheduling observations of agile satellites. Aerosp Sci Technol 2002;6(5):367–81.

6. Wolfe WJ, Sorensen SE. Three scheduling algorithms applied to the earth observing systems domain. Manag Sci 2000;46(1): 148–66.
7. Chen Y, Zhang D, Zhou M. Multi-satellite observation schedule algorithm based on hybrid genetic particle swarm optimizationAdvances in information technology and industry applications. Berlin, Heidelberg: Springer; 2012. p. 441–8.
8. Mao T, Xu Z, Hou R. Efficient satellite scheduling based on improved vector evaluated genetic algorithm. J Netw 2012;7(3): 517–23.
9. Xhafa F, Sun J, Barolli A. Genetic algorithms for satellite scheduling problems. Mobile Inf Syst 2012;8(4):351–77.
10. Xhafa F, Herrero X, Barolli A. Evaluation of struggle strategy in genetic algorithms for ground stations scheduling problem. J Comput Syst Sci 2013;79(7):1086–100.
11. Xhafa F, Sun J, Barolli A, et al. Evaluation of genetic algorithms for single ground station scheduling problem. Advanced information networking and applications (AINA), 2012 IEEE 26th international conference on 2012 Mar 26–29; Fukuoka, Japan; 2012. p. 299–306.
12. Globus A, Crawford J, Lohn J, Pryor A. Scheduling earth observing satellites with evolutionary algorithms. In: Conference on space mission challenges for information technology (SMC-IT), 2003 July; Pasadena, California, USA; 2003. p. 1–7. http:// people.nas.nasa.gov/~globus/papers/SMCIT03/ SMCIT02 paper3. pdf.
13. Globus A, Crawford J, Lohn J, Morris R. Scheduling earth observing fleets using evolutionary algorithms: problem description and approach. In: Conference on Space Mission Challenges for
1. Information Technology (SMC-IT); 2003. Pasadena, California, USA July 13-16, 2003. p. 27–29. http://people.nas.nasa.gov/ ~globus/papers/SMCIT03/ paper1.pdf
14. Gao P, Tan YJ, Li JF. An ant colony algorithm for remote satellite and ground integration scheduling problem in parallel environment. Adv Mater Res 2013;791:1341–6.
15. Feng MY, Tang SX, He J. Resource scheduling for space-based early warning system based on improved particle swarm optimization. J China Acad Electron Inf Technol 2010;5(1):97–101.
16. Bianchessi N, Cordeau JF, Desrosiers J, Gilbert L, Vincent R. A heuristic for the multi-satellite, multi-orbit and multi-user management of earth observation satellites. Eur J Oper Res 2007;177(2):750–62.
17. Sarkheyli A, Vaghei BG, Bagheri A. New tabu search heuristic in scheduling earth observation satellites. Software technology and engineering (ICSTE), 2010 2nd international conference on, 2010 Oct 3–5; San Juan PR; 2010. p.V2-199–V2-203.
18. Marinelli F, Nocella S, Rossi F. A Lagrangian heuristic for satellite range scheduling with resource constraints. Comput Oper Res 2011;38(11):1572–83.

19. Frank J, Jonsson A, Morris R, Smith D. Planning and scheduling for fleets of earth observing satellites. In: Proceedings of the sixth international symposium on artificial intelligence, robotics, automation and space, Montreal, Canada, June 18-21, 2001. http:// ti.arc.nasa.gov/m/profile/de2smith/publications/iSAIRAS01-EOSscheduling.pdf.
20. Jufang L, Feng Y, Baocun B, He RJ. A decomposition-based algorithm for imaging satellites scheduling problem. In: International conference on information engineering and computer science (ICIECS); 2009 Dec 19–20; Wuhan, China: 2009. p. 1–6. http:// ieeexplore.ieee.org/xpls/abs_all.jsp?arnumber=5363469&tag=1.
21. Li JF, Tan YJ. Modeling and solving observation satellites scheduling as a pickup and delivery problem. Syst Eng Theory Pract 2004;12:65–71.
22. Zhu KJ, Li JF, Baoyin HX. Satellite scheduling considering maximum observation coverage time and minimum orbital transfer fuel cost. Acta Astronaut 2010;66(1):220–9.
23. Hao C, Jun L, Ning J, Liu XG, Tang Y. Scheduling model and algorithms for autonomous electromagnetic detection satellites. Acta Aeronaut Astronaut Sin 2010;31(5):1045–53.
24. Hao C, Jun L, Ning J, Tang Y. The model and algorithm of dynamic scheduling for electromagnetic detection satellite adapted to task change. Signal Process 2009;25(11):1659–65.
25. Marco D, Vittorio M, Alberto C. The ant system: optimization by a colony of cooperating agent. IEEE Trans Syst Man Cybern B Cybern 1996;26(1):29–41.

CITATION

Yuqing Li, Rixin Wang, Minqiang Xu, Rescheduling of observing spacecraft using fuzzy neural network and ant colony algorithm, Chinese Journal of Aeronautics, Volume 27, Issue 3, June 2014, Pages 678-687, ISSN 1000-9361, http://dx.doi.org/10.1016/j.cja.2014.04.027.

Chapter 4

The Space Mission Design Example Using Leo Bolos

Oleg Nizhnik

ISAS/JAXA, Yoshinodai 3-1-1, Chuuou-ku, Sagamihara-shi 252-5210, Kanagawa-ken, Japan

ABSTRACT

Four sample space launch missions were designed using rotating momentum transfer tethers (bolos) within low Earth orbit and a previously unknown phenomenon of "aerospinning" was identified and simulated. The momentum transfer tethers were found to be only marginally more efficient than the use of chemical rocket boosters. Insufficient power density of modern spacecraft's was identified as the principal inhibitory factor for tether usage as a means of launch-assistance, with power densities at least 10 W/kg required for effective bolos operation.

INTRODUCTION: PREVIOUS RESEARCH ON MOMENTUM-TRANSFER TETHERS

After 56 years of space exploration, the space launch industry remains reliant only on chemical rockets. This paper will investigate if any merit exists in implementing an alternative launch assist technology with functionality of existing upper stages—momentum transfer tethers.

For this purpose, Section 1 reviews the literature on the tether-assisted propulsion. Section 2 quantifies how specifics of the LEO (low-earth orbit) environment—a relatively dense atmosphere with a small density scale—along with micrometeoroids will affect operations of the tethers. Section 3 discusses the demerit of implementing tip masses or counterweights on momentum-transfer tether. Section 4 reviews the literature on the proper calculation of the tether tensile strength. Section 5 reviews existing solutions for the payload capture using tethers, and an original solution, based on a mid-air satellite retrieval system is proposed. In Section 6, four example orbital launch missions with momentum-transfer tethers are designed and their parameters analyzed. Section 7 concludes on the areas of development most critical for future implementation of the momentum-transfer tethers as launch assistance means.

In 2003, Frisbee [1] stated that the necessity to create a sustainable (high launch-rate) infrastructure was historically the delay to both launch-assist catapults and tether-assisted propulsion. The concept of an earth-based catapult typically has multi-GW electrical power requirements, due the necessity of launching crafts heavy enough to pierce the Earth's atmosphere. In 2013, Powell, with the StarTram team [2], promoted a design with peak electrical power of 104 GW, and that power cannot be scaled down to any reasonable (10–100 MW) range. However, momentum-transfer tethers operate at near-vacuum and can be implemented even with currently meager energy technology. However, how high must the launch rate be? Do the recent advances in ion propulsion and high-strength polymers improve the merit of momentum transfer tethers? Are rotating tethers stable? Let's try to answer these questions, starting with a review of already completed research related to space tethers.

Experiments were performed in space to master tether deployment. The work of Gates, in 2001 [3], stresses the importance of fast (>1 m/s) deployment to avoid problems with built-in stresses and electric charges. The longest tether stably deployed to date has reached a length of 31.7 km, albeit unintentionally [4]. According to Aerospace America Magazine [5], in 2010, the near-term target was to achieve reliable deployment and operation of 1 km tethers, although other projects [6] succeeded with non-librating deployment of 20 km-long tethers in 1993. In 2007, Menon [7] described in detail a tether brake used to control deployment of the 31.7 km-long tether in the SpaceMail experiment [4].

Therefore, the author concludes that current technology is sufficient to deploy 10–20 km-long tethers in the LEO environment.

Although significant research has been conducted by the scientific community on hanging or librating tethers for momentum transfer, in 1888, Wiesel [8] proved that optimal orbital transfer was achieved using a tether with constant tension during the entire transfer process. Therefore, all hanging and librating tether configurations were marked as less efficient, in terms of momentum transfer, for a given tether mass.

Among many rotating momentum-transfer tether configurations promoted on the Tether Unlimited site, the LEO bolo (rotating tether) [9,10] deserves special attention, as it is scalable to low tip speed. The orbital geometry of the LEO bolo is shown in Figure 1. The tether of the LEO bolo is also reusable; although, in 2010, Williams [11] proposed using a system with two rotating tethers for orbital capture—one disposable and one reusable. However, non-reusable tethers must be orders-of-magnitude stronger than existing ropes to be mass-effective as means of momentum transfer.

The LEO bolo can even operate the propellant-free, extracting momentum necessary for orbital transfer from the momentum of bodies in higher orbits. Unfortunately, around the Earth, the only such body is the moon, and series of tethers able to handle at least 1.6 km/s tip speed are necessary for harnessing the orbital momentum of the moon rocks. An example of momentum-neutral tether design may be found in [12]. Even for sub-km/s delta-v offered by bolo configurations built with existing materials, a MXER (magnetic-exchange electrodynamic reboost) drive option [10] still exists. But MXER is highly speculative research; Sorensen, in 2003 [10], lists several components of the MXER bolo as having a NASA TRL (technology readiness level) equal to 3 (only proof-of concept available), and two key technologies (tether coating and tether libration control) having a TRL equal to 2 (only technology concept was formulated). In addition, a high degree of difficulty (III) in R&D is expected for many components. Therefore, alternative engineering approaches to MXER should be considered first. The author favors modern ion motors as near-term alternatives to momentum-neutral or MXER systems.

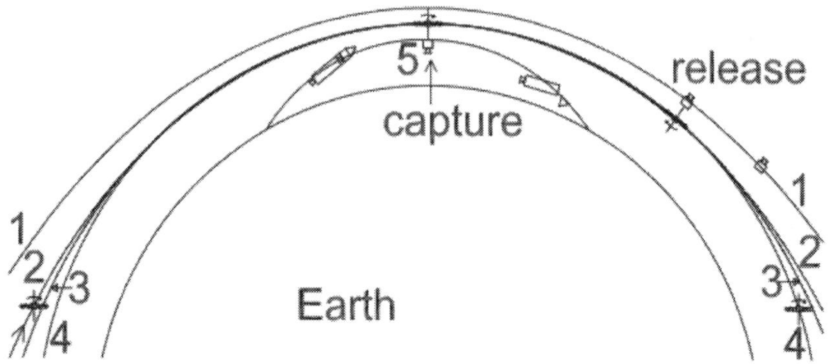

Figure 1. Orbital geometry of the LEO bolo (rotovator) launch assistance mission. (1) payload post-release orbit; (2) tether pre-capture orbit; (3) tether and payload post-capture orbit; (4) tether post-release orbit (circular); (5) payload pre-capture sub-orbital trajectory.

The problem of rotating bolo (rotovator) stability has been extensively discussed by the research community. In 1996, Luo [13] described an extreme wealth of rotating tether vibration types. Such vibrations may easily result in tether straying from the correct capture or release point, or may even result in tether breakage. Therefore, two parametric measures should be taken. Firstly, a material with low accumulation of elastic energy (*i.e.*, high rigidity) should be used for the tether. Secondly, a damper should be used as most rigid materials themselves have very low internal mechanical energy dissipation. In 1995, Puig-Suari [14] stated that, even after implementation of critical dampening, peak load due to tether vibrations exceeds steady-state load by 50%. Of course, aero-capture, as in [14], is the worst case, in terms of vibration excitation, but some vibration will be excited in the LEO bolo, as well, with each dip of the tether arm in the deeper layers of atmosphere. Even changes of tether temperature due to entering sunlight or shadow may result in vibration, which ultimately must be damped. In 1995, Tyc [15] speculated that large flexible booms attached to the tether, OEDIPUS-A, acted as a surprisingly effective dampening mechanism because of their high-frequency, high-strain whipping oscillations induced by the oscillations of the tether. In 1997, Vigneron [16] proposed and experimentally verified the stability criteria for a rotating tether holding a payload. His findings can be summarized as a statement of vibration instability of any tether that is too long or rotating too slowly. Both, faster rotation and smaller length contribute to larger natural dampening factors. In 1991, Matteis [17] demonstrated that tether dampening is rather ineffective in suppressing fundamental transverse (libration) mode (approximately a period of 400 s for an 80 km tether), although modes with higher frequency, including longitudinal vibration modes, can be effectively suppressed. Therefore, it is important for a tether system to avoid periodic forcing at or near fundamental transverse mode frequency. Such periodic forcing may happen due to atmospheric interaction as soon as the tether tip dips into the more dense atmospheric layers. In 2003, Takeuchi [18] stated that although librations of uncontrolled tethers on elliptical orbits are inevitable, with periodic thrust applied on the tether tips, these librations can be suppressed in cases of massless tethers. In addition, Takeuchi [18] found the steady-state rotating tether solution similar to that described in Section 2, but did not quantify it. By a superposition principle, stabilization of a librating massive tether on an elliptical orbit may require a periodic thrust distributed across the length of the tether. In 1987, Martinez-Sanchez [19] demonstrated that tether stability in an orbital plane may be maintained by reeling/unreeling the tether as well.

Therefore, the author concludes that LEO bolo may rotate (but not liberate) predictably if excessive lengths (beyond 100 km) are not considered.

THE LEO ENVIRONMENT: DRAG, MOMENTUM BUILDUP (AEROSPINNING), AND METEOROIDS

This section complements the technical designs of momentum transfer tethers in Section 1 with the discussion on physical specifics of the tether operations in the LEO environment. Tether lifetime and tether rotation speed stability will be quantified in the current Section.

Engineers accustomed to the LEO mission design are familiar with the phenomenon of air drag. Air drag impacts much of the lifetime and requirements for propulsion for all LEO spacecrafts. However, often overlooked, is the rotation momentum imparted on the LEO tethers by air drag.

The rotation momentum buildup (aerospinning) phenomenon is defined as self-sustained and self-limited prograde rotation of an object in orbit due to its interaction with the density gradient of the atmosphere. Intuitively, rotation torque appears as the lower portion of the tether encounters denser air. Therefore, the center of pressure is initially below the tether's center of mass. The diagram of forces resulting in aerospinning is shown in Figure 2. The smaller atmosphere density scale and longer tethers will result in the rotation momentum becoming significant. Such a mechanism always rotates a tether in prograde direction, resulting in the lower parts of the tether having progressively lower air velocities. As tether rotation speed increases, the center of pressure moves up until equilibrium rotation speed is reached with the center of pressure coincident with the center of mass. The unsteady process before reaching the equilibrium rotation speed may include libration and variable-speed rotation (as described by Takeuchi in [18]) and to call this "spin-up" is proposed by the author.

The role of atmosphere in tether angular movement did not go unnoticed previously. In 1949 and 1972, Philips [20] and Stevens [21], respectively, stated the theoretical possibility of self-sustained transverse oscillations in tensioned tethers, if tangential airspeed exceeds the speed of the transverse waves in the tether. As a solution, Stevens [21] recommended providing separation between fundamental longitudinal and transverse frequencies to avoid large-amplitude cross-coupled oscillations induced by airflow. In 2004, Jokic [22] simulated the spin-up of tethers due to interaction with the planetary atmosphere, although only for transient interaction during the aerocapture. In 1997, Puis-Suari [23] also demonstrated that tethers may be rotated by the transient interaction with the atmosphere. Overall, members of the research community to date have acknowledged the possibility of tether rotation initiation because of the interaction with the airflow, but failed to see the self-limiting nature of such a rotation, and the principal role of the atmospheric density gradient in initiating and sustaining a rotation.

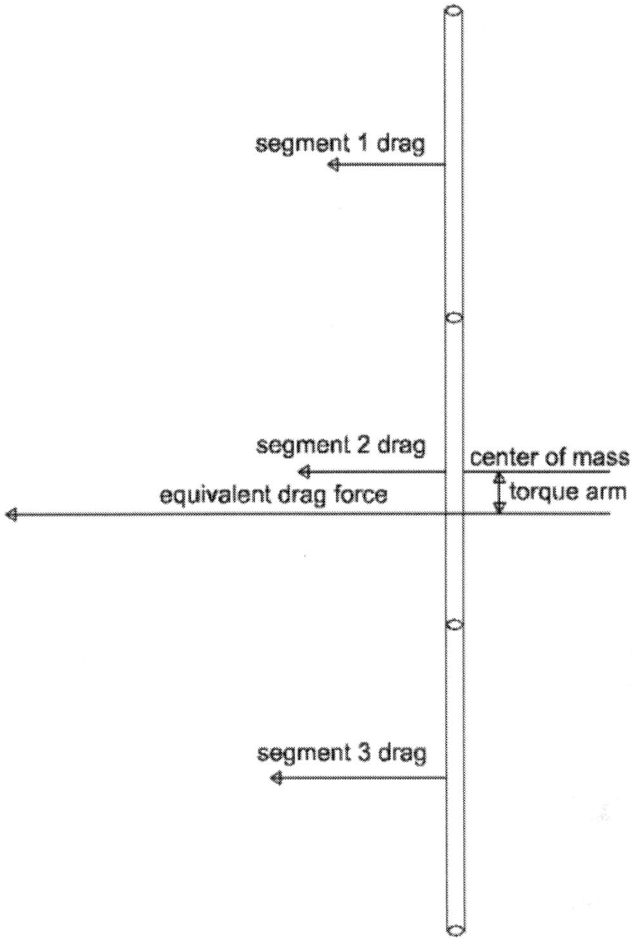

Figure 2. The diagram of forces resulting in tether aerospinning.

Both, air drag and air-induced torque, are dependent on the ratio between spacecraft (tether) unit mass and cross-sectional area. For unity surface divided by mass, both, drag and momentum have an associated characteristic time. These are: time to reentry for air drag, and steady angular speed divided by initial angular acceleration for air-induced momentum (called a characteristic spin-up time in Figure 5).

The author has simulated normalized (1 kg weight and 1 m^2 ram area tethers) reentry and spin-up times using the numerical integration in conjunction with the MSISE-90 atmospheric model. The simulation results of reentry time and aerospinning steady-state tip speed are presented in Figure 3 and Figure 4, respectively. For the lower atmosphere (below 400 km) it was found that the characteristic spin-up time to reentry time ratio is close to 1000, but tends to decrease in the upper atmosphere. Such a large "drag-to-spin" ratio indicates what spin-up phenomenon will manifest itself only after several orbital re-boosts, if re-boosting force is applied to the center of mass. Hence, it is possible to cancel aerospinning by slightly offsetting the re-boost engine thrust axis from the tether center of mass, although at the cost of added engineering complexity.

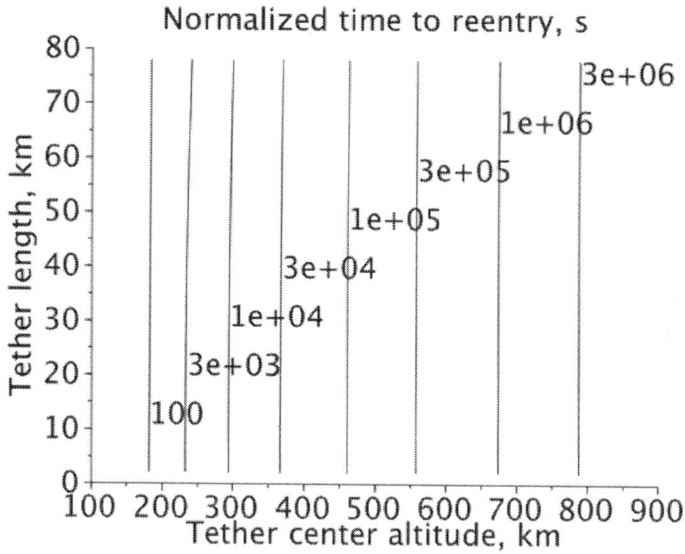

Figure 3. The normalized reentry time for tethers on equatorial circular orbits.

As the numerical integration algorithm may suffer from high sensitivity to input parameter variations, the orbital lifetime model was verified against Table I-1 in [24]. No discrepancy was found for small tether lengths, indicating general stability of reentry (and therefore, air drag) numerical integration. For longer tethers, at altitudes below 200 km, a braking effect of the tether tip entering dense layers of atmosphere become noticeable, while, for higher altitudes, drag was only slightly dependent on tether length because of the longer atmospheric density scale in the thermosphere. Additional drag is rather small as, for normalized tethers, additional drag in inhomogeneous atmospheres is the second-order correction to the drag of compact surfaces. This additional drag is visible in Figure 3 as a slight curvature of the reentry time contours below 3000 s of normalized reentry time.

Figure 4 shows the steady-state speed of the tether tip for average solar activity. Solar activity greatly changes the density scale of the upper atmosphere. Therefore, the indicated speeds can vary as much as three times those at low solar activity (resulting in greater speeds at low orbits) and high solar activity (resulting in lower speeds at low orbits). However, as the molar weight of the upper atmosphere is increased during the solar maximum, an altitude range (550–750 km) exist such that the effects of increased temperature and increased molar weight roughly cancel each other, resulting in a more stable atmospheric density scale and, therefore, a relatively stable steady-state rotation speed.

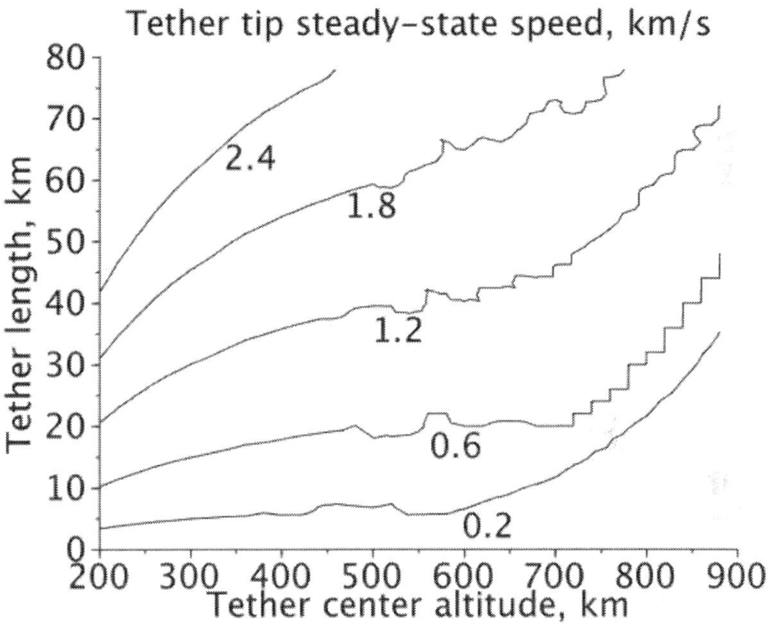

Figure 4. Steady-state tip speed of the rotating tether on the equatorial circular orbit in average solar conditions (based on MSISE-90 atmospheric model).

The important feature visible on Figure 5 is the range of orbit altitudes and tether lengths, whereby the steady-state rotation speed of the tether is nearly independent of the phase of the solar cycle. At this range, the tethers will rotate in a steady-state, with a period of roughly 170 s. Due to the large (compared to the Solar cycle period) spin-up time scale, it is expected that tethers on the rotation stability line will experience variability of their steady-state tip speed below 10%.

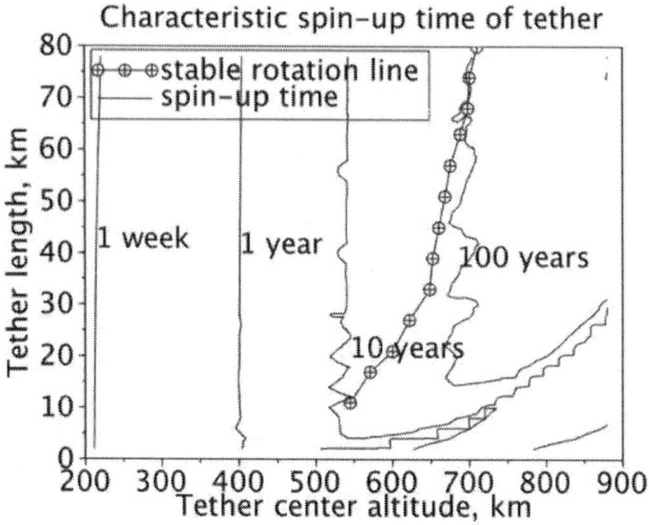

Figure 5. The characteristic spin-up time of tethers on equatorial circular orbits.

Due to the gravity gradient stabilization, tether rotation cannot begin spontaneously as deployed. An initial tether angular velocity must be supplied by the momentum wheel or tip jet to start the rotation.

To calculate the rotating tether tip minimal speed at the nadir (or zenith) point, it is possible to equate the rotational energy of the tether at this point to the tidal energy of the tether stretched vertically (Equation (1a)). Equation (1a) states that to sustain rotation, the tether must have enough rotational energy to librate from a stable vertical orientation to an unstable horizontal orientation, with inertia of the tether acting against the tidal forces. Solving Equation (1a) with the gravity potential of Earth yields Equation (1b).

$$\frac{I_{tether}\omega_{nadir}^2}{2} = E_{gravity}(horizontal\_orientation) - E_{gravity}(vertical\_orientation) \quad (1a)$$

$$\upsilon_{\min} = \sqrt{\left(\frac{L}{R_{earth} + h_c} + \ln\frac{R_{earth} + h_c - L/2}{R_{earth} + h_c + L/2}\right) \cdot \frac{12GM_{earth}}{L}} \tag{1b}$$

where L represents the length of the tether; h_c, the tether orbit altitude; R_{earth}, the radius of Earth; G, the gravity constant; M_{earth}, mass of Earth.

A plot of minimal initial tip speed of a hanging tether is shown in Figure 6.

In 2005, Pelaez [25] stated that a stationary solution for the tether tip argument in elliptic orbits does not exist, and a tether tip must librate as the gravity vector is rotating non-uniformly. Therefore, for elliptic orbits, the minimal rotation speed is reduced compared to circular orbits. This finding was also partially supported by a study of Takeuchi, the results of which were published in 2003 [18].

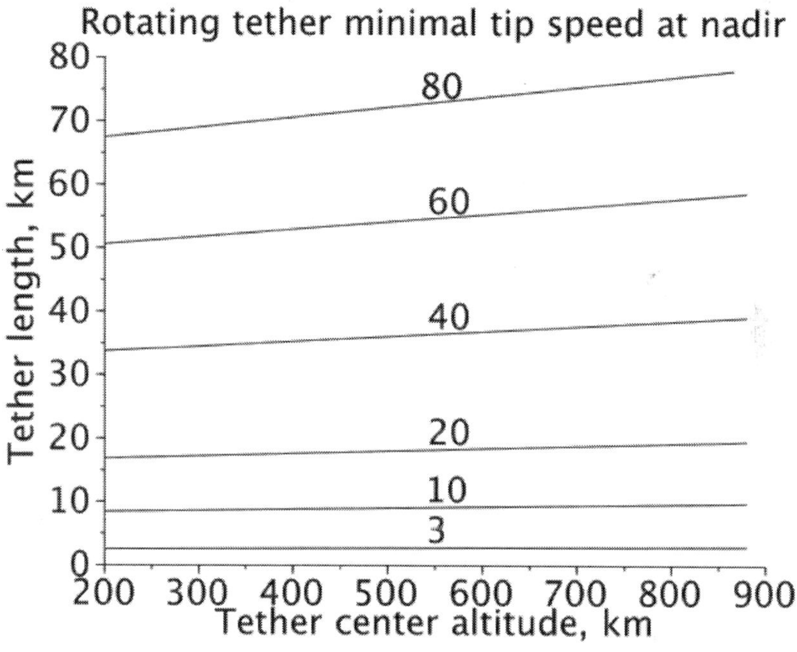

Figure 6. The minimal nadir speed of the tether tip relative to the tether center of mass to sustain tether rotation.

Micrometeoroids and orbital debris (MMOD) hazards play an important role in tether design.

The colliding objects frequency on LEO can be extracted from Figures 7–19 in [24] and approximated by Equation (2) for debris ranging from 4 μm to 2 cm in size. The collision probabilities are as summarized in Table 1.

$$N_{year} = 1.6 \cdot 10^{-10} D^{-2.56} \quad (2)$$

Table 1. Collision frequencies for various debris sizes.

Particle size	Collision frequency for 1 m² single-side target
100 μm	2.8 collisions/year
200 μm	0.48 collisions/year
500 μm	0.045 collisions/year
1 mm	0.0077 collisions/year
2 mm	0.0013 collisions/year
2.5 mm	0.00074 collisions/year
3 mm	0.00046 collisions/year

Although, in 1995, Forward and Hoyt [26] claimed a redundancy of the fail-safe tether (hoytethers) may be implemented without sacrificing the tensile strength, it is highly unlikely, given the extreme difficulty of producing lightweight coupling to the fibers with high tensile strength. Couplings in damaged hoytethers would subject redundant filaments to the compressive stress. All high-strength fibers are weak in compression (with typical tensile-to-compression strength ratios being 10–40 [27]. Fatigue strength of filaments may be degraded as well, because zones of compression are a likely place for cracks to nucleate. Therefore, contrary to the conclusion of Forward and Hoyt [26], the failures of individual filaments in single hoytether segments may be correlated. As such, stated increases of redundant tether (hoytether) lifetimes may be not achievable, although further experiments are necessary to validate or reject such a claim.

An alternative, an advanced form of a Whipple shield (stopping 6.3 mm particles for 10.8 kg/m^2 weight already flown [28]) may be better suitable for higher-diameter tethers. The stuffed Whipple shield (multishock shield) model with multiple layers spaced by 1.8 mm and a weight of 0.4 kg/m^2 per layer was integrated into software used to generate the sample mission budgets in Section 6.

THE INCONVENIENCE OF THE TIP MASSES OR COUNTERWEIGHTS

This Section is intended to outline the conditions when the momentum transfer tether with tip mass or counterweight may be useful. In addition, the related problem of matching rotation phase of the bolo with the payload trajectory will be discussed with a resulting conclusion on the inconvenience of tip masses or counterweights for launch-assist missions using LEO bolos.

In 2001, Ruiz [29] proposed to use momentum-transfer tethers for orbit apogee raising. Ruiz [29] found that additional tip mass raises the transfer orbit apogee, but no tether reuse, spin-up, or catch requirements were considered.

The bolo must not only be deployed to LEO, but also rotated to its operational tip velocity if aero-spinning alone cannot be relied upon. Such a spin-up requirement is non-trivial, and requires, most realistically, a hydrazine-powered (or electric-ion) low-thrust rocket pod on either end of the tether. Therefore, minimization of the energy required for spin-up will allow reduction of bolo launch mass. Such minimization of energy spent generally requires minimization of the moment of inertia of the entire bolo. Therefore, placement of any massive components on the bolo tip (as counterweight) will result in larger energy requirements for bolo spin-up, as illustrated in Figure 7. Any energy spent for bolo spin-up is lost for momentum transfer because payload is only subject to radial acceleration by the bolo.

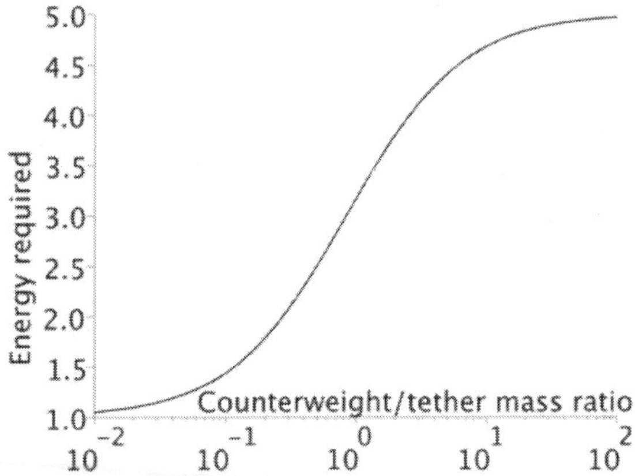

Figure 7. The energy (normalized) spent to spin-up a tether with a point-mass counterweight on one end, from standstill to fixed free tip velocity.

In addition, it is advisable to reduce the bolo moment of inertia to allow easy adjustment of the rotation phase. Such an adjustment is necessary to catch payloads launched from the fixed Earth's location to the lowest point of tether.

Relying on a booster rocket to provide cross-time to meet the bolo tip in its lower point will result in the booster flying long segments in non-ballistic trajectory, with resulting delta-v losses of 400–800 m/s.

The phase of the bolo can be adjusted either by tip rockets (very expensive, in terms of propellant mass and added weight) or by reeling/unreeling the tether to the central hub. In 2006, Williams [30] devised a method to change the spin rate of the rotovator (bolo) using Coriolis forces on the tether in a yo-yo configuration, without any reaction thrusters. The reeling/unreeling mechanism provides necessary cross-time capability for 1–10 W power levels (250 kg payload designs in Section 6) and reel speed below 1 mm/s. Maximal reeled length is 100 m for 43 s cross-time compensation within a two-day reeling maneuver. In 1998, Tragesser [31] also proposed adjusting the aerodynamic drag of the tether with planetary atmosphere during flyby, by means of winding/unwinding the tether.

Given the lesser role of aerospinning (described in Section 2) at altitudes above 300 km and the necessity to minimize the moment of inertia of bolo, the optimal bolo configuration would be the symmetric tether with powered spacecraft for bolo re-boost placed at or near the tether center. Counterweight may still be considered if heavy MMOD protection or the reduction of aero-spinning spin-up rate (for very low orbits) is important.

TETHER MATERIALS

This Section links the momentum transfer tether system architecture and environment discussion inSection 1, Section 2 and Section 3 with the mission design in Section 6 by providing guidelines as well as examples to the calculation of the physical properties of the tethers. In particular, tether strength and maintenance are discussed.

The primary component of the bolo is the high-tension rope (tether), characterized by the critical tip velocity, dependent on strength to weight ratio. In 2001, Ziegler [32] calculated an equation for the maximal permissible rotation speed (or maximal payload for given rotation speed) of tether with a payload attached to one end (See Equation (3a,b)).

$$\omega_{\max} = \sqrt{\frac{F_{break}}{L(m_p + m_t/2)}} \quad (3a)$$

$$m_p = \frac{F_{break}}{\omega^2 L} - m_t/2 \quad (3b)$$

However, using only ultimate tensile strength and density of material to calculate breaking force may lead to over-optimistic calculations of critical tip velocity and tether mass, as happened to Jokic [33], as the strength of the tether (rope) is mostly a function of defects present in tether material, not the material. For example, although the highest reported measured strength of microscopic alumina fibers was 22.3 GPa [34], the same material from the same manufacturer has a strength of 6.7 GPa [27] if made into long (0.5 m) fibers. Such a decrease of strength is a combination of the effect of the easy slip planes, surface imperfections, and voids inside the alumina fibers. In 2002, Elises [27] published a detailed review of high-strength fibers and their failure modes, and any tethered-mission designer is strongly advised to read it.

The following derating from the ultimate tensile strength should be used:

1) If loads are periodic or constant (as opposed to impulsive), use maximal yield strength;
2) If brittle failure is possible in metal elements by a factor of 3 [27];
3) For time-dependent degradation (especially for Zylon);
4) For a maximal operating temperature (or up-rate, if operating temperature is low);
5) For ultraviolet light and radiation-induced degradation;
6) For the bundling of yarn (not all yarns are tensioned equally)—this penalty increases with tether diameter;
7) For the weaving of rope (bundles are twisted and pre-stressed);
8) For atomic oxygen erosion (for tethers on LEO) or decomposition catalysts present (copper on polyethylene);
9) For the creep during service life (most important for polymer ropes, especially UHMWPE) [35];

10) For the extra length of the tether compared to length of test specimens (variability of strength increases logarithmically with an increase in sample length).

All derating coefficients may be cumulative.

Typically, after all deratings, loading all of the potential tether materials beyond 30% of their respective ultimate tensile strength is dangerous, and some synthetic fibers (UHMWPE, polyester, Kevlar in marine environments) may be usable only up to 5%–20% of their ultimate tensile strength [27].

One is sometimes misled by the stated structural factors of 1.5 to 2.0 (corresponding to loads 67% to 50% of ultimate tensile strength) of the structural components (wings, for example) of aircraft. Such low structural factors are possible only if component is available for defectoscopy or is flying a very short time. The impressive reliability of modern aircrafts is the function not only of advanced materials, but also of inspection infrastructure. If one cannot monitor (and as necessary, repair) small cracks or scars before they grow to a dangerous size, it is better to double the structural factors of component in question.

In addition, the geometry of the rotating tether is very unfavorable to timely crack detection, even if defectoscopy equipment is available during normal tether operation. Critical cracks may be only 1–2 mm long in steel fibers (necking and then brittle failure) and 10–20 mm long in UHMWPE ropes (creep, fibrillar, and then brittle failure).
For example, let us estimate usable 10 km tether strength for two materials: 2 GPa New Japan Steel wire for suspension bridge cable [36] and Dyneema SK78 0.54 mm multifilament UHMWPE yarn [35]. The comparison is provided in Table 2.

Due to the complex nature of derating, as much data as possible was derived from the actual testing of the ropes rather than from material specifications.

Overall, although UHMWPE provide 10× better average strength-to-weight ratio compared to steel filament, the performance of the

UHMWPE is more derated because of strength variability, insufficient testing, and creep.

Table 2. The usable strength of the example high-tension materials.

Materials	Akashi bridge cable wire	Dyneeema SK78 multi-filament yarn
Ultimate tensile strength	2.0 GPa	2.34 GPa (for a fill factor of 0.78)
Diameter	5.23 mm	0.54 mm × 10
Effective density	7700 kg/m^3	756 kg/m^3 (for a fill factor of 0.78)
Length to implement	10 km	10 km
(1) Yield derating	IIMD *	IIMD *
(2) Brittleness derating	0% (below fatigue limit)	0% (below creep limit)
(3) Degradation derating	IIMD *	n/a
(4) Temperature derating	IIMD *	IIMD *
(5) Radiation derating	0% (steel is radiation hard)	30% (estimated)
(6) Bundling derating	n/a (single wire)	IIMD * for 0.54 mm, 12% for 5.4 mm
(7) Weaving derating	n/a (single wire)	IIMD *
(8) Erosion derating	0.0007%/year (274 km orbit)	0.26%/year (274 km orbit)
(9) Creep derating	IIMD *	IIMD * for 1 year lifetime (−70%) #
(10) Length derating	0% (tested in 2 km pieces)	−70% (7% strength variability in 0.5 m test samples up-rated to 10 km length)
Total IIMD * derating	−68.75%	70%
Total derating (1 year)	−68.75%	81.6%
1/(Structural factor)	IIMD *	33.3%
Usable tether strength	625 MPa &	287 MPa
Derated tip velocity	284 m/s	616 m/s
Tether usable strength	13.4 kH	15.0 kH

Tapered tethers, although theoretically more mass-effective, were not considered. Although the tapered fibers are known to naturally occur in sea urchins and sea cucumbers, which condones the superior strength of their ligaments, as noted in 2000 by Trotter [37], engineering processes for the manufacturing of tapered fibers or ropes is absent. The only available option to imitate a tapered tether is hence a connection of the ropes of different thicknesses through the adapter. Such adapters impose additional weight and reliability penalties; therefore, adapters are practical only for large tapering ratios.

THE CHALLENGE OF THE BOLO RENDEZVOUS AND CATCHING

This Section defines the interface between highly speculative momentum transfer architecture and hardware outlined in Section 1, Section 2, Section 3 and Section 4 and existing rocket propulsion hardware. A literature review on payload capture and tentative overview of hardware components for payload capture will be provided in this section.

Catch of payload by LEO bolo has very strict requirements for speed, location, and time-matching between the payload and bolo tip. A space-docking experience is not applicable for rendezvous with momentum transfer tethers. In 2006, Williams [38] investigated the low-speed tether capture dynamic. A conclusion was that, to avoid transient loads, the capture time must be at least two orbital revolutions—an unsuitably long interval for bolo tip capture. For high-speed rendezvous, an asteroid grapple was proposed in 2000 by Lanoix [39] for a tethered swing-by of the asteroids to modify the spacecraft's speed vector. Such an approach is intrinsically unreliable and far beyond the capability of the existing throwing grapples.

Historically, most resembling the bolo capture was the air recovery of Corona satellites, Stardust sample recovery, or attempted recovery of the Genesis sample capsule. A satellite being recovered closely resembles the bolo tip, while recovery aircraft mostly resemble the suborbital, maneuvering payload. The Corona satellite was captured at high relative velocities (up to 60 m/s) using a combination of hooks hanging from aircrafts and loops created by parachute ropes attached to the Corona satellite. It is a flight-proven solution.

Research is currently in progress, specifically for bolo by Newton [40] at the Tennessee Tech University, but only on room-scale, open-loop models. The grappling aperture of Newton [40] is heavy, weighing 10% of the payload mass. Furthermore, a capture happens at zero relative velocity—a feature requiring multi-second speed adjustment phase in the actual flown, close-loop controlled payload. The speeds of the bolo tip and payload cannot be matched far away from the rendezvous point because of the natural uncertainties of the payload trajectory. Most of the uncertainty happens due to interaction with winds, turbulence, unsteady rocket engine thrust, fuel slosh, and rocket stage pointing errors. It is common for the last stage of the orbital vehicle to be left within ~10 m/s of the desired speed, even if bi-propellant liquid-fuel engines are used. Solid-fuel engines have an even worse final speed spread (Pegasus XL official 3-sigma error is 150 m/s [41]). Therefore, a multiple correction maneuvers must occur at the last moments, guided most likely by bolo tip radio beacon, GPS, and laser rangefinder. Even as such, intervals of a few seconds of the bolo tip and payload at low relative speed (<30 m/s) are desirable to roughly adjust range and orientation errors, given the typical low agility of the modern orbital boosters.

Requirements for the orbital booster (or payload only) to have an increased agility in order to be caught will surely result in payload-bearing modules having too much deadweight, out-weighing any benefit from any Δv savings provided by the bolo.

Given these constraints, the only currently achievable flight profile leading to successful bolo rendezvous is reuse of the main engine in the last stage of the orbital booster to roughly match the acceleration of the bolo tip for a few seconds before the catch moment.

Typical upper stages have linear accelerations up to 30 m/s^2, throttleable down to 60% of nominal thrust, and have angular accelerations up to 4 rad/s^2 (typical for Falcon-1e upper stage). To equalize longitudinal and transverse agility, these parameters result in free acceleration of 6 m/s^2 with acceleration lagging about 0.2 s after controls. An approximate correctable course deviation as a function of the rendezvous window duration can be approximated as Equation (4):

$$\Delta r = \frac{(a_{\max} - a_{\min})(t_{rendezvous} - N_{corrections}\ t_{lag})^2}{4} \qquad (4)$$

If $N_{corrections}$ =4, t_{lag} =0.2 s, $a_{max} - a_{min}$ =12 m/s^2, $t_{rendezvous}$ =3 s, the correctable trajectory error is 15 m, which is within the error margin provided by the commercial GPS. But the same trajectory correction maneuver will cost the upper stage a Δv proportional to the rendezvous window duration, which is expressed as Equation (5).

$$\Delta \upsilon_{loss} = \frac{a_{\max} - a_{\min}}{2}(t_{rendezvous} + t_{engine\_startup}) \qquad (5)$$

Such a Δv loss for terminal guidance would be in the order of 100 m/s for the typical upper-stages and a rendezvous window duration of 3 s.

Payloads with attachment hardware must be designed to disconnect from last stage immediately on successful attachment with the tether tip, and at the same moment the last-stage engine must start the shutdown to prevent re-contact with the payload. Engine shutdown speed requirements are expected to be less severe compared to normal stage separation due to the payload acceleration provided by the tether.

In Figure 8, a tentative concept of payload capture-associated hardware for LEO bolo is shown. The concept is derived from the Corona satellite recovery scheme and explained in detail below.

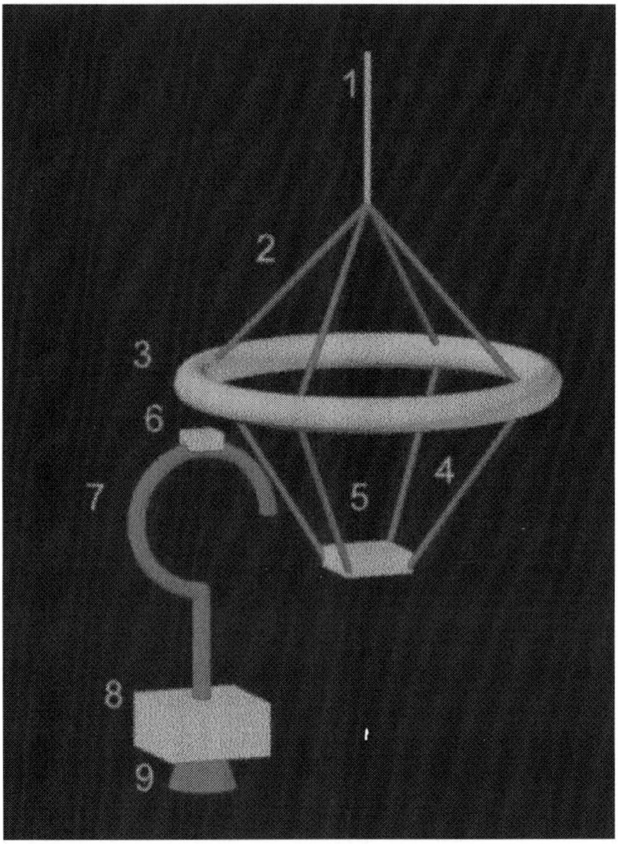

Figure 8. A tentative hook-and-loop catch mechanism for LEO bolos. (1) tether; (2) stiffening rods; (3) load redistribution ring; (4) shock absorbing ropes; (5) saddle for hook; (6) hook release (pyrotechnic-actuated hinge); (7) payload hook; (8) payload; (9) payload adapter to rocket last stage.

Tether design (1) is covered in Section 4 of this paper. The most optimal construction of stiffening rods (2) and load redistribution rings (3) is high-stiffness Li-Al alloy pipes. Given the ultimate tensile strength of such constructs 550 MPa, with a structural factor of 8 (typical for shock-loaded structures) and ring diameter of 3 m, parts (2) and (3) matching the strength of tether calculated in Table 2, can be constructed of 15 m Li-Al pipes weighing 7 kg in total. The most suitable material for shock absorbing ropes (4) is polyester, widely used for similar roles in rock climbing. Using the same analogy as rock climbing, the saddle (5) should be made of high-toughness steel, totaling 5 kg for (4) and (5) by up-scaling normal rock climbing gear to the loads in Table 3. The pyrotechnic bolt (6) or other actuator is necessary to provide a rapid payload release, and must have a weight below 0.5 kg. The 3-m-long payload hook (7) is the most weight-critical component because it must be launched with each payload, and also requires increased payload fairing size. Although the hook itself will be about 2 kg (less than 1% of payload mass as calculated in Section 6), the added weight and size of fairing may increase gravity loss and air drag Δv requirements. Therefore, such a hook system strongly favors booster design without the need of fairing. Such designs were proposed in 2012 by Nizhnik [42]. The payload (8) must be strengthened to accommodate additional stress of tether catch, but because in the proposed configuration, the main shock is aligned with normal rocket booster thrust vector, payload weight penalty is likely going to be negligible. The payload adapter (9) of the Ariane-5 rocket, as described in [43], is completely adequate for the tether catch operation. Therefore, no further development is necessary on payload adapter.

Table 3. The space launch mission budgets using momentum transfer tethers (bolo) on LEO.

Mode	Existing steel tether	Existing PE tether	10 W/kg hub and PE	Double-strength PE
Catch mass	246 kg	247 kg	247 kg	243 kg
Tether rotation period	52 s	73 s.	73 s	152 s
Tether length	4 km	9 km	9 km	32 km
Tether tip speed	242 m/s	387 m/s	387 m/s	661 m/s
Tether tip acceleration	29.2 m/s^2	33.3 m/s^2	33.3 m/s^2	27.3 m/s^2
Tether tension without payload	6.2 kH	6.8 kH	6.8 kH	23.3 kH
Tether tension with payload	13.4 kH	15 kH	15 kH	30 kH
Δv for payload	474 m/s	759 m/s	691 m/s	1310 m/s
Δv for tether [&]	13.8 m/s	16.4 m/s	88.7 m/s	15.3 m/s
Tether mass (with shield)	852 kg	810 kg	810 kg	3416 kg
Tether MMOD shield weight	190 kg	651 kg	651 kg	3257 kg
Tether MMOD shield blocking class	2 mm	2.5 mm	2.5 mm	3 mm
Tether effective drag area	20.6 m^2	56.7 m^2	56.7 m^2	238 m^2
Total effective drag area (night glider mode)	30.3 m^2	70.3 m^2	70.1 m^2	260.5 m^2
Tether hub power	7.6 kW	10.7 kW	11.2 kW	17.4 kW
Tether hub mass	7.6 ton	10.7 ton	1.1 ton	17.4 ton
Tether catch orbit, perigee × apogee (km)	299 × 321	357 × 383	322 × 458	420 × 444
Post-release (circular) tether orbit altitude	299 km*	357 km*	322 km*	420 km*
Pre-capture payload orbit, perigee × apogee (km)	111 × 299	−218 × 365	−214 × 317	−591 × 404
Post-release payload orbit, perigee × apogee (km)	301 × 487	375 × 958	327 × 858	448 × 1455
Payload increase [#]	25 kg	42 kg	34 kg	79 kg
Base payload	246 kg	247 kg	247 kg	243 kg
Tether hub thrust for post-release orbit departure	60 mH	45 mH	90 mH	53 mH
Thrust necessary for tether 1 week-long re-boost	193 mH	311 mH	283 mH	527 mH
Refueling frequency, every Nth payload	N = 24	N = 17	N = 17	N = 10
Lifetime required to break-even mass on LEO	10.89 years	7.95 years	1.91 years	7.32 years
Average time to break due micrometeorites	115 years	97 years	97 years	44 years

An alternative solution may be to spin a secondary tether around the tip of the primary tether, nullifying tether tip acceleration at the moment of capture and therefore allowing low-acceleration rendezvous. That solution was first proposed in 2004 by Williams [44], but a dedicated research program may be necessary to develop it.

EXAMPLE MISSIONS DESIGNS

In this section, the conclusions on momentum transfer architecture (Section 1 and Section 3), environment (Section 2), tether hardware (Section 4), and tether interface (payload capture, Section 5) are combined in a coherent LEO constellation delivery mission. This mission is compared against a similar mission, which uses only chemical propulsion—namely, Falcon-1e vehicles. To avoid potentially misleading economic and payload utilization considerations, mission merit is determined as mass of hardware placed on at least a 185 km LEO orbit by using equal numbers of Falcon 1e launches. Fractional launches (part of the rocket-delivering part of the payload) are also considered possible for simplicity.

A software package was developed in the Scilab for the semi-automatic calculation of the mass budget of the space launch mission using momentum transfer tethers. The Scilab scripts are provided on request or available on the author's page at www.researchgate.net. A total of nine scripts were used and briefly described below. The mission budgets derived from auto-budget script outputs are provided in Table 3.

The Tether_reentry.sce script works as follows

a) The MSISE-90 atmospheric model is loaded and interpolated with a resolution of 1 km.
b) The distribution of atmospheric drag across the tether is calculated at five segments across the tether length for rotation angles with step 15 degrees.
c) The circular orbit altitude drop is calculated from energy and impulse conservation laws. Spacecraft assumed to collide inelastically with air molecules, therefore transferring the kinetic and potential energy of spacecraft to air, accelerating and ultimately heating it. The application of conservation laws is possible if the orbit decaying quasi-statically. Quasi-staticity mean drag acceleration several orders of magnitude below gravity acceleration, and gravity acceleration magnitude varying negligible per orbit. For this to happen, orbit must be near-circular and gravity field must be nearly-spherical. Real decaying low-earth orbits are slightly non-circular (orbital eccentricity is varying around 0.001 due non-spherical gravity fields). In addition, a few minutes before reentry, decay becomes not quasi-static due to increasing drag acceleration.
d) The time step is adapted to keep the orbit altitude drop rate in the 0.1–1 km/step range.
e) Simulation is terminated after the orbit is lowered to 100 km altitude.
f) The steps b–e are swept for various tether lengths and initial orbit altitudes. The tether rotation speed is derived from the Steady_tether_speed.sci script output.

The Steady_tether_speed.sci script works as follows:

a) The MSISE-90 atmospheric model is loaded and interpolated with a resolution of 1 km.
b) Initially assumed tip speed is 1 m/s and the speed adaptation coefficient is 1.4.
c) Torques arising from center of pressure and center of mass mismatch are calculated for nine tether segments and averaged over a rotation period with step 3 degrees.
d) If torque is found to be positive, estimated speed is multiplied by speed adaptation coefficient until torque becomes negative.
e) If torque is found to be negative, estimated speed is divided by speed adaptation coefficient until torque becomes positive.
f) When torque changes sign, the speed adaptation coefficient is set to the square root of itself.
g) The steps c–f are repeated until torque absolute value is reduced below the pre-determined accuracy.
h) The steps b–g are repeated while sweeping orbit altitude and tether mass.

The Tether_spinup_time.sci script works as follows:

a) The MSISE-90 atmospheric model is loaded and interpolated with a resolution of 1 km.
b) The output of Steady_tether_speed.sci is divided by torque calculated for zero rotation speed, but averaged for all possible tether orientations with step 3 degrees.

The tether_minspeed.sce script simply applies Equation (1) for the range of the tether lengths and circular orbit altitudes.

The tether_mmod.sce script calculates the total weight of MMOD shield for a given length and diameter of tether, using tabulated MMOD shield area density extracted from [28].

The tether.sce script uses laws of energy and momentum conservation to calculate the following parameters as a function of payload weight to tether weight ratio:

a) The Δv imparted on the tether as a result of payload capture and release.
b) The Δv imparted on the payload after being carried by the tether from a capture point directly below the tether hub to a release point directly above the tether hub.

The auto-budget scripts, (Tether_autobudget_UHMWPE.sce and Tether_autobudgetSteel.sce), do automatically calculate power and propellant requirements for the catch-release-reboost momentum transfer cycle, given average orbit altitude and launch frequency. Optimization of these is left for the user. Other script outputs are used as input for auto-budget scripts. Average altitude of the tether capture orbit was swept to minimize orbital mass break-even time. Break-even time is the interval necessary for the bolo system to transfer, to orbit, additional (compared to purely rocket system) mass equal to the initial bolo weight (plus all ion motor fuel spent in the process). It was discovered that higher launch rates result in the reduction of orbital break-even time, up to a launch frequency of 180 per year. It was also found that a launch frequency above 52 launches/year (weekly) results in a bolo system that is undeliverable to orbit by medium-lift vehicles. Therefore, only bolo capture orbit average altitude was optimized, while launch frequency was fixed to 52 launches/year. For more robust optimization convergence, the gravity gradient on LEO was linearized. Therefore, all apogee and perigee data in Table 3 are approximations only. The tether strength data was derived from the Table 2. Typical satellite power density (1 W/kg) and solar panel energy output (150 W/m^2) were used for columns 1, 2, and 4. Solar panels are assumed to have an effective drag factor of 1 in sunlight and 0.2 in shadow, corresponding to the "night glider" operation mode.

The payload catch is assumed to happen near the perigee of the bolo orbit to avoid perigee lowering and excessive air drag during re-boost. The payload is released after a half-rotation of the bolo, still near the bolo perigee, to maximize momentum transfer (in line with [45]). After payload release, the tether orbit is considered circularized.

In 2006, Ahedo [46] explained how any conductivity of the tether on Earth's orbit will result in electromagnetic force, causing eventual tether de-orbit. Therefore, only insulating or semi-insulating tethers (with insulator junctions between segments) can be considered for long-term, low-loss operation. However, for sake of simplicity, that effect for steel (hence, conductive) tethers was not taken into account. Surprisingly, the concept that the re-boost engine specific impulse is a limiting factor, as stated in 2003 by Sorensen [10], was proven wrong, as fuel spent for currently existing ion motors (1500 s ISP and 50% efficiency) was found to reduce tether-assisted payload increase for about 1/4 only. This conclusion is in line with analysis by Sanmartin [47], who rejected magnetic-reboosted (MXER) tethers. Sanmartin [47] states that a tether-magnetic system will be too heavy, compared to the electric propulsion, if orbital transfer times below one month are required due to poorer power efficiency and associated weight of the necessary photovoltaic panels.

In addition, a small role of the material strength was found. Although bolo performance (pay-off time) moderately improves (by 27%) if steel tether is replaced for available high-strength UHMWPE rope (five times stronger for same weight), further doubling of tether strength results only in a 9% improvement in pay-off time. The main reason for such a performance stagnation is the limiting role of the payload catch specifications. If acceleration at the tether tip is kept at 30 m/s^2, as advocated in Section 5, a faster tether tip speed results in an increase of tether length, with an associated increase of the MMOD shield weight. In addition, high-speed bolos tend to be most efficient at higher altitudes because of a large drag-to-mass ratio. Therefore, although Δv delivered may be impressive (1.31 km for double-strength UHMWPE rope), most of the delivered Δv is over the Δv required for LEO orbit insertion. It must be stressed that all Δv beyond minimal-altitude (185 km approximately) circular LEO delivered by bolo system is simpler (and cheaper) to deliver using onboard ion motors or inter-orbital ion tug, for example, the ion tug designed by Jaffe [48] or Hermel [49]. Ion tug will clearly be more efficient in the absence of the heavy, damage-prone tether. In 1972 and 1975, Mason [50] and Strauss [51], respectively, questioned the usefulness of the ion tugs compared to expendable chemical propulsion. Absence of the ion tugs, as recently as 2013, has proven their statement.

However, power density of the bolo hub spacecraft was found to be very important, with 4.16 times the break-even time reduction if power density rises from 1 W/kg to 10 W/kg. In addition, although total Δv delivered to payload is reduced because of the lower bolo mass, Δv to reach catch point is actually decreased because of the bolo lower optimal orbit altitude. The entire Δv penalty for the light (high power-density) hub is, thus, applied for reduction of the apogee of the post-release orbit. The result is more circular post-release orbits compared to missions using standard power density spacecrafts. These orbits are well-matching the launch-assistance purpose of the LEO bolo, outlined in Section 1

CONCLUSIONS

This section summarizes the factors critical for the efficacy of the LEO launch assistance mission using LEO bolos, calculated in Section 6. Five directions for making such missions feasible are outlined, with most promising direction being the increase of spacecraft power/weight ratio.

The development of a hypothetical space mission involving a satellite constellation delivered to orbit with the help of a LEO bolo has resulted in the following conclusions:

Specific impulse (~1500 s) of modern ion motors is sufficient to deliver a 3:1 reduction of fuel mass spent for bolo-assisted momentum transfer (compared to chemical rockets). Development of MXER tethers is not critical for the momentum-transfer applications, unless tethers with tip speeds approaching 1 km/s will be deployed.

Long-term strength of existing ropes is sufficient for LEO bolos. Tether maintenance crawler(s) may offer significant reduction of tether strength margin(s), but increasing tether strength does not significantly improve system performance, due to the added weight of the tether re-boost hardware and micrometeoroid shield.

The power density of modern spacecrafts (1 W/kg) is inadequate for LEO bolos deployment. Development of higher power-density (up to 10 W of electrical power per kg) spacecrafts is necessary if realistic (~1–40 launches/year) launch rates are expected. Otherwise, launch rates of 50–200 launches/year are required to pay-off the initial cost of LEO bolo deployment (even if not taking into account R&D costs)

The power-to-drag of modern solar arrays in "night glider" mode is inadequate for bolos operating below a 300 km altitude. Better power efficiency of solar cells or solar panels operating in "sun slicer" mode are desirable. However, operating in "sun slicer" mode still requires many advances in large structures and energy storage. A nuclear powered bolo for a 240–280 km catch orbit may be considered if politically acceptable.

Plain tethers are vulnerable to micrometeoroid damage. Either shielded or redundant tethers must be used. Currently, shielded tethers seem to be the most developed solution, and are weight-effective for tethers below 10 km long.

Payload capture requirements strongly favor high-agility upper stages of rocket boosters to reduce terminal guidance Δv loss.

ACKNOWLEDGMENTS

This work was partially supported by Japan Science and Technology Agency within the ERATO (Exploratory Research for Advanced Technology) project.

CONFLICTS OF INTEREST

The author declares no conflict of interest.

REFERENCES

1. Frisbee, R.H. Advanced space propulsion for the 21st century. *AIAA J. Propuls. Power* **2003**, *19*, 1129–1154.
2. Powell, J.; Maise, G.; Rather, J. Maglev Launch: Ultra Low Cost Ultra/High Volume Access to Space for Cargo and Humans. Available online: http://www.startram.com/resources (accessed on 17 December 2013).
3. Gates, S.S.; Koss, S.M.; Zedd, M.F. Advanced tether experiment deployment failure. *AIAA J. Spacecr. Rocket.* **2001**, *38*, 60–68.
4. Kruijff, M.; van der Heide, E.J. Data analysis of a tethered SpaceMail experiment. *AIAA J. Spacecr. Rocket.* **2009**, *46*, 1272–1287. [

5. Tethers in Space. AIAA Aerospace America Magazine, December 2010, 59–64.
6. Lorenzini, E.C.; Bortolami, S.B. Control and flight performance of tethered satellite small expendable deployment system-II. *AIAA J. Guid. Control Dyn.* **1996**, *19*, 1148–1156.
7. Menon, C. Design and testing of a space mechanism for tether deployment. *AIAA J. Spacecr. Rocket.* **2007**, *44*, 927–939.
8. Wiesel, W.E. Optimal payload lifting with tethers. *AIAA J. Guid.* **1988**, *11*, 352–356.
9. Hoyt, R.P. Responsive a Launch of Small Spacecraft Using Reusable in-Space Tether and Air-Launch Technologies. Available online: http://www.tethers.com/papers/MXERSpace2006Paper.pdf (acces sed on 18 December 2013).
10. Sorensen, K. Momentum Exchange Electrodynamic Reboost (MXER) Tether Technology Assessment Group Final Report. 2003. Available online: http://www.nasaspaceflight.com/_docs/MXER%20TAG%202003% 20Report.pdf (accessed on 18 December 2013).
11. Williams, P. Tether capture and momentum exchange from hyperbolic orbits. *AIAA J. Spacecr. Rockets* **2010**, *47*, 205–209.
12. Hoyt, R.P. Cislunar tether transport system. *J. Spacecr. Rocket.* **2000**, *37*, 177–186.
13. Luo, A.C.J.; Han, R.P.S.; Tyc, G. Analytical vibration and resonant motion of a stretched spinning nonlinear tether. *AIAA J. Guid. Control Dyn.* **1996**, *19*, 1162–1171.
14. Puig-Suari, J.; Longuski, J.M.; Tragesser, S.G. Aerocapture with a flexible tether. *AIAA J. Guid. Control Dyn.* **1995**, *18*, 1305–1312.
15. Tyc, G. Attitude dynamics investigation of the OEDIPUS-A tethered rocket payload. *AIAA J. Spacecr. Rocket.* **1995**, *32*, 133–141.
16. Vigneron, F.R.; Jablonsky, A.M.; Chandrashaker, R. Comparison of analytical modeling of OEDIPUS tethers with data from tether laboratory. *AIAA J. Guid. Control Dyn.* **1997**, *20*, 471–478.
17. De Matteis, G.; de Socio, L.M. Dynamics of a tethered satellite subjected to aerodynamic forces.*AIAA J. Guid.* **1991**, *14*, 1129–1135.
18. Takeuchi, N.; Natori, M.C.; Okuizumi, N. Fundamental strategies for control of a tethered system in elliptical orbits. *AIAA J. Spacecr. Rocket.* **2003**, *40*, 119–125.
19. Martinez-Sanchez, M.; Gavit, S.A. Orbital modifications using forced tether-length variations.*AIAA J. Guid.* **1987**, *10*, 233–241.
20. Phillips, W.H. Theoretical Analysis of the Oscillation of a Towed Cable. Available online: http://naca.central.cranfield.ac.uk/reports/1949/naca-tn-1796.pdf (accessed on 18 December 2013).
21. Stevens, G.W.H. A theory of vibrations in parachutes. *AIAA J. Aircr.* **1972**, *9*, 74–78.

22. Jokic, M.D.; Daniel, W.J.T. Aerogravity-assist maneuvering of a tethered satellite system. *AIAA J. Spacecr. Rocket.* **2004**, *41*, 614–621.
23. Piug-Suari, J. Optimal mass flexible tethers for aerobracking maneuvers. *AIAA J. Guid. Control Dyn.* **1997**, *20*, 1018–1024.
24. Wertz, J.R.; Everett, D.F.; Puschell, J.J. *Space Mission Engineering: The New SMAD*; Microcosm Press: Portland, OR, USA, 2011.
25. Pelaez, J.; Andres, Y.N. Dynamic stability of elecrodynamic tethers in inclined elliptical orbits.*AIAA J. Guid. Control Dyn.* **2005**, *28*, 611–622.
26. Forward, R.L.; Hoyt, R.P. Failsafe Multiline Hoytether Lifetimes. In Proceedings of the 31st AIAA/ASME/SAE/ASEE Joint Propulsion Conference and Exhibit, San Diego, CA, USA, 10–12 July 1995. AIAA Paper 95–28903.
27. *Fiber Fracture*; Elises, M., Llorca, J., Eds.; Elsevier: Oxford, UK, 2002.
28. Dye, S. MMOD-IMLI: Thermal Insulation and Micrometeoroid/Orbital Debris Protection. NASA SBIR/STTR Technologies, SBIR Phase I Proposal. Available online: http://ehb8.gsfc.nasa.gov/sbir/docs/public/recent_elections/SBIR_10_P1/SBIR_10_P1_104617/briefchart.pdf (accessed on 16 April 2013).
29. Ruiz, M.; Pelaez, J. Influence of a small end mass on tether-mediated orbital injection. *AIAA J. Guid. Control Dyn.* **2001**, *24*, 1108–1117.
30. Williams, P. Optimal deployment/retrival of a tethered formation spinning in the orbital plane.*AIAA J. Spacecr. Rocket.* **2006**, *43*, 638–650.
31. Tragesser, S.G.; Longuski, J.M. Analysis and design of aerocapture tether with accounting for stochastic errors. *AIAA J. Spacecr. Rocket.* **1998**, *35*, 683–689.
32. Ziegler, S.W.; Cartmell, M.P. Using motorized tethers for payload orbital transfer. *AIAA J. Spacecr. Rocket.* **2001**, *38*, 905–914.
33. Jokic, M.D.; Longuski, J.M. Design of tether sling for human transportation systems between earth and mars. *AIAA J. Spacecr. Rocket.* **2004**, *41*, 1010–1015.
34. Soltis, P.J. Anisotropy of the tensile properties of submicron-size sapphire (Al_2O_3) whiskers. *Bull. Am. Phys. Soc.* **1965**, *10*, 163.
35. Chimisso, F.E.G. *Some Experimental Results Regarding Creep Behavior on Synthetic Materials Used to Produce Offshore Morring Ropes*; Mechanics of Solids: Rio Grande, Brazil, 2009.
36. Honshu-Shikoku Bridge Expressway Company Limited, 2013. Available online: http://www.jb-honshi.co.jp/technology/material.html (accessed on 13 December 2013).
37. Trotter, J.A.; Tipper, J.; Lyons-Levy, G.; Chino, K.; Heuer, A.H.; Liu, Z.; Mrksich, M.; Hodneland, C.; Dillmore, W.S.; Koob, T.J.; *et al.* Towards a fibrous composite with dynamically controlled stiffness: Lessons from echinoderms. *Trans. Biochem. Soc.* **2000**, *28*, 357–362.
38. Williams, P. In-plane payload capture with an elastic tether. *AIAA J. Guid. Control Dyn.* **2006**, *29*, 810–821.

39. Lanoix, E.L.M.; Misra, A.K. Near-earth asteroid missions using tether sling shot assist. *AIAAJ. Spacecr. Rocket.* **2000**, *37*, 475–480.
40. Newton, K. NASA Engineers, Tennessee College Students Successfully Demonstrate Catch Mechanism for Future Space Tether. Available online: http://www.nasa.gov/centers/marshall/news/news/releases/2005/05–108.html (accessed on 15 April 2013).
41. Isakowitz, S.J.; Hopkins, J.B.; Hopkins, J.B., Jr. *International Reference Guide to Space Launch Systems*, 4th ed.; AIAA: Reston, VA, USA, 2004; p. 295.
42. Nizhnik, O. A low-cost launch assistance system for orbital launch vehicles. *Int. J. Aerosp. Eng.***2012**, *2012*, 1–10.
43. Fortesque, P.; Swinerd, G.; Stark, J. *Spacecraft Systems Engineering*, 4th ed.; Wiley: Hoboken, NJ, USA, 2011; p. 254.
44. Williams, P.; Blanksby, C. Prolonged payload rendezvous using a tether actuator mass. *AIAA J. Spacecr. Rocket.* **2004**, *41*, 889–893.
45. Kyroudis, G.A. Advantages of tether release of satellites from elliptic orbits. *AIAA J. Guid.* **1988**,*11*, 441–448.
46. Ahedo, E.; Sanmartin, J.R. Analysis of bare-tether systems for deorbiting low-earth-orbit satellites.*AIAA J. Spacecr. Rocket.* **2002**, *39*, 198–205.
47. Sanmartin, J.R. Efficiency of electrodynamic tether thrusters. *AIAA J. Spacecr. Rocket.* **2006**, *43*, 659–666.
48. Jaffe, L.D. Nuclear-electric reusable orbital transfer vehicle. *AIAA J. Spacecr. Rocket.* **1988**, *25*, 375–381.
49. Hermel, J.; Meese, R.A.; Rogers, W.P.; Kushida, R.O.; Beattie, J.R.; Hyman, J. Modular, ion-propelled, orbit-transfer vehicle. *AIAA J. Spacecr. Rocket.* **1988**, *25*, 368–374.
50. Mason, J.D. Space tug performance optimization. *AIAA J. Spacecr.* **1972**, *9*, 491–492.
51. Strauss, E.L. Ablative thermal protection for space tug multipass and aerobraking entry. *AIAA J. Spacecr. Rocket.* **1975**, *12*, 346–350.

CITATION

Oleg Nizhnik, The Space Mission Design Example Using LEO Bolos, doi:10.3390/aerospace1010031.

Chapter 5

On-Orbit Camera Misalignment Estimation Framework and Its Application to Earth Observation Satellite

Seungwoo Lee and Dongseok Shin

Satrec Initiative, 21 Yusung-daero 1628 Beon-gil, Yuseong-gu, Daejeon 305-811, Korea;

ABSTRACT

Despite the efforts for precise alignment of imaging sensors and attitude sensors before launch, the accuracy of pre-launch alignment is limited. The misalignment between attitude frame and camera frame is especially important as it is related to the localization error of the spacecraft, which is one of the essential factors of satellite image quality. In this paper, a framework for camera misalignment estimation is presented with its application to a high-resolution earth-observation satellite—Deimos-2. The framework intends to provide a solution for estimation and correction of the camera misalignment of a spacecraft, covering image acquisition planning to mathematical solution of camera misalignment. Considerations for effective image acquisition planning to obtain reliable results are discussed, followed by a detailed description on a practical method for extracting many GCPs automatically using reference ortho-photos. Patterns of localization errors that commonly occur due to the camera misalignment are also investigated. A mathematical model for camera misalignment estimation is described comprehensively. The results of simulation experiments showing the validity and accuracy of the misalignment estimation model are provided. The proposed framework was applied to Deimos-2. The real-world data and results from Deimos-2 are presented.

INTRODUCTION

The alignments of sensors and actuators inside a spacecraft are measured carefully in laboratory before launch. The alignment between the attitude frame and the camera frame is called boresight alignment [1]. Since it is essential for the quality of earth-obs ervation satellites [2], several studies have already developed various methods for the precise alignment in laboratory [3,4,5].

Despite the efforts to the precise alignment before launch, on-orbit calibration is mandatory because additional errors may be introduced after launch by launch shock, outgassing, zero-gravity, thermal effect,*etc.* Therefore, it is common for earth-observation satellite programs to calibrate the alignment during the initial commissioning period [2,3,6,7,8,9,10,11,12,13,14,15,16,17,18,19,20]. However, it is difficult to obtain an accurate alignment because the alignment error is measured as the sum of various error sources [6].

A common approach for on-orbit boresight alignment calibration is based on the usage of ground control points (GCPs) and a physical sensor model. Breton and Bouillon used multiple GCPs and a physical model of the spacecraft to estimate the alignment of SPOT-5 [7,8]. Radhadevi also used a similar approach for IRS-P6 [6] which requires as many data samples as possible in order to obtain reliable results. However, the number of GCPs hardly exceeded hundreds because their GCPs relied on spots or landmarks prepared beforehand [6,7,8,9,10,11].

Some researchers tried an automatic GCP extraction technique using reference ortho-photos and image-based feature matching to get a much higher number of GCPs [12,13,14,21,22]. Robertson *et al.* even included the automated GCP extraction step to RapidEye's ground processing chain for alignment calibration [14]. However, the alignment was calculated for individual images, and the system's alignment was simply taken from the average of individual measurements. This approach can cause biased measurement in case that GCPs are distributed unevenly in an image. It is also unable to consider different alignment characteristics of attitude sensors. Müller *et al.* proposed a processing chain for automated geo-referencing that includes automatic GCP extraction and sensor model improvement [21]. They successfully geo-referenced thousands of images from SPOT-4, SPOT-5, and IRS-P6 using automatically extracted GCPs. However, their study used thousands of images that were already taken, and they did not cover the considerations for effective on-orbit calibration. Klančar *et al.* suggested an image-based attitude control mechanism that correlates the spacecraft camera image and the reference image on the fly [22]. Although this approach eliminates the need of measuring the boresight alignment, it is impractical due to the limited on-board resource to have a high-resolution GCP database on-board and perform the real-time feature matching.

Several interesting approaches were proposed for Pleiades-HR, such as single-track reverse imaging and star imaging, as well as the GCP method [16,17,18]. The single-track reverse imaging method, which is called auto-reverse, utilizes the spacecraft's high agility to rotate the spacecraft 180 degrees after imaging and to take a second image of the same spot [16]. This method looks quite promising and has several strengths over the GCP method. However, it is only applicable to high performance satellites such as Pleiades-HR. The star-imaging method is an operationally efficient method which could be performed during the eclipse period without interfering daylight imaging operation [17]. It is, however, still in a conceptual stage and not developed yet for practical applications. It also has a potential risk to measure different or unwanted error, as space imaging is different from ground imaging. Despite the proposal of new calibration methods, the baseline method for alignment calibration of Pleiades-HR was the GCP-based method [18] which used only 20 GCPs on average per site from 20 spots across the world.

The latest research on alignment calibration concerns reports on ZY1-02C [19,20]. Wang *et al.*provided a well-written explanation of the interior and exterior orientation error determination and an estimation model that uses many GCPs extracted from an ortho-photo [19]. Although it is essential to use multiple images to estimate misalignment from noisy data, they used a single image for estimation in the experiment. The acquisition planning to build proper estimation dataset is also important as the coverage of spacecraft attitude and attitude sensors affect the accuracy and robustness of the estimation; however, they did not consider this aspect. Jiang *et al.* tried to correct dynamic attitude error by correlating overlapped CCD images [20]. In their article, the GCPs were collected from ortho-photo manually, which led them to a small number of available GCPs for experiment.

Camera misalignment is the misalignment between the attitude frame and the true camera frame. The camera frame is calculated from the spacecraft attitude and pre-launch alignment measurement data, whereas the true camera frame is where the boresight vector is actually pointing. The static error between those two frames is observed as if the spacecraft has a biased error on its attitude. The boresight alignment calibration could be done by finding the camera misalignment and compensating it from the spacecraft attitude.

In this paper, a framework for on-orbit camera misalignment estimation of earth-observation satellites is presented. It provides an all-in-one solution from the planning of ground-target image acquisition to the estimation of the camera misalignment. Some of the important aspects when choosing the calibration targets are discussed, as well as the distribution of tilt angles and sensor selections for effective calibration. A proven robust automatic approach for the extraction of many GCPs from ortho-photos is explained in detail. Investigating the pattern of localization errors of GCPs is essential. Sometimes misalignments of attitude sensors are not the same, generating different error patterns depending on which set of attitude sensors were used. The images must be grouped by parameters showing similar error patterns (e.g., sensors, imaging modes, target locations), and the bias must be estimated separately for each group. The mathematical model, which was derived from colinearity to solve the camera misalignment, is explained comprehensively. Simulation tests with sophisticated simulation setup are conducted to prove validity of the framework. Finally, the application results to the Deimos-2 camera misalignment estimation during the initial commissioning phase are presented to show the effectiveness of the proposed framework.

In Section 2, the overview of the camera misalignment estimation is described, followed by the necessary background knowledge using Deimos-2 as an example. The steps of camera misalignment estimation such as image planning, automated GCP extraction, localization error pattern analysis, and the mathematical model for camera misalignment estimation are also explained. Section 3 describes the results of the experiments to evaluate the proposed framework. The accuracy of the mathematical model was evaluated using simulation data. The results of the application to Deimos-2 are demonstrated as well. In Section 4, the summary of the work is presented.

CAMERA MISALIGNMENT ESTIMATION FRAMEWORK

Overview

Conceptually, the estimation process of camera misalignment consists of four major steps. The first step is planning the satellite's imaging operation to acquire image data for the camera misalignment estimation. Extracting GCPs from the images is the next step. Analyzing the error pattern of GCPs and confirming the existence of misalignment are followed. The last step is to estimate the camera misalignment using the GCPs and spacecraft ancillary, which contains spacecraft position and attitude data. Figure 1 shows the conceptual process of the camera misalignment estimation.

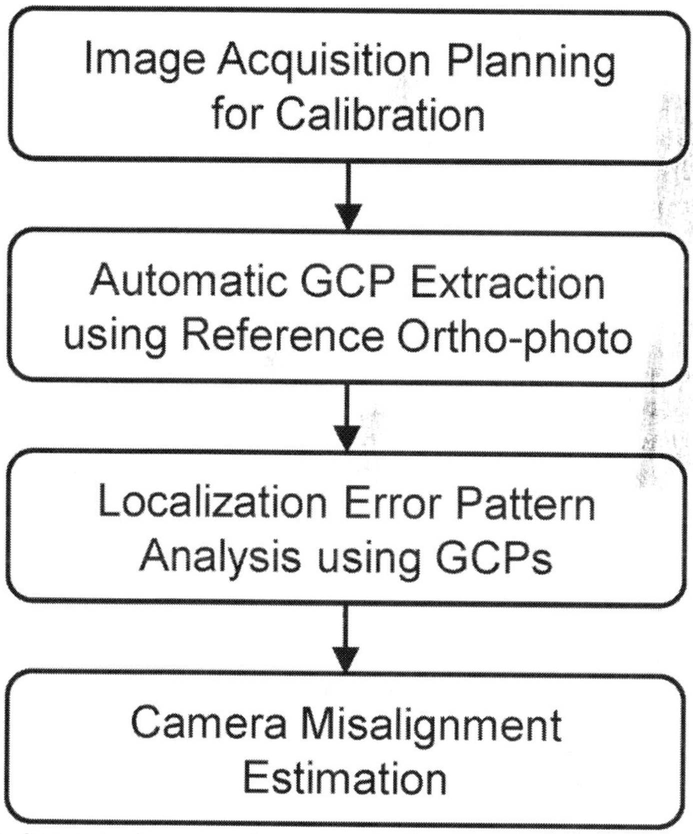

Figure 1. Conceptual process of camera misalignment estimation.

Whereas the simple four steps are provided from a conceptual view, the structural view of camera misalignment estimation involves a couple of additional steps as illustrated in Figure 2.

After planning imaging scenarios for calibration, the image collection planning system (ICPS) uploads the plans to the spacecraft (1). After the spacecraft takes images, the images are downloaded to the ground station from the spacecraft (2); thereafter, the image receiving and processing system (IRPS) generates ortho-image products, which consist of geometrically corrected images and spacecraft ancillary data (3). AutoGCP software correlates the images with a reference ortho-photo database and generates GCPs and sensor vectors (4). Camera Misalignment Estimator estimates the camera misalignment by using the results of AutoGCP and spacecraft ancillary data (5). The estimated camera misalignment is then set to IRPS in order to adjust the sensor model (6).

It is recommended to iterate the process (3)–(6) at least a couple of times, because large camera misalignment can make the IRPS use a wrong location for the DEM for geometric correction, which influences the accuracy of the calculated geo-location. Therefore, the camera misalignment needs to be re-estimated after applying the misalignment estimate. The iteration may stop when the change of misalignment estimate is less than the desired accuracy.

Background: Deimos-2

The proposed framework was applied to Deimos-2 which is a high-resolution earth observation satellite equipped with push-broom type CCD sensors providing 1.0 m resolution panchromatic band (PAN) and 4.0 m resolution multispectral band images (blue, green, red, and near-infrared) with 12 km swath width. Since the estimation is done using attitude measurement and image GCPs, it is important to understand the target system's camera geometry, attitude determination mechanism, and the definitions of attitude and camera frames.

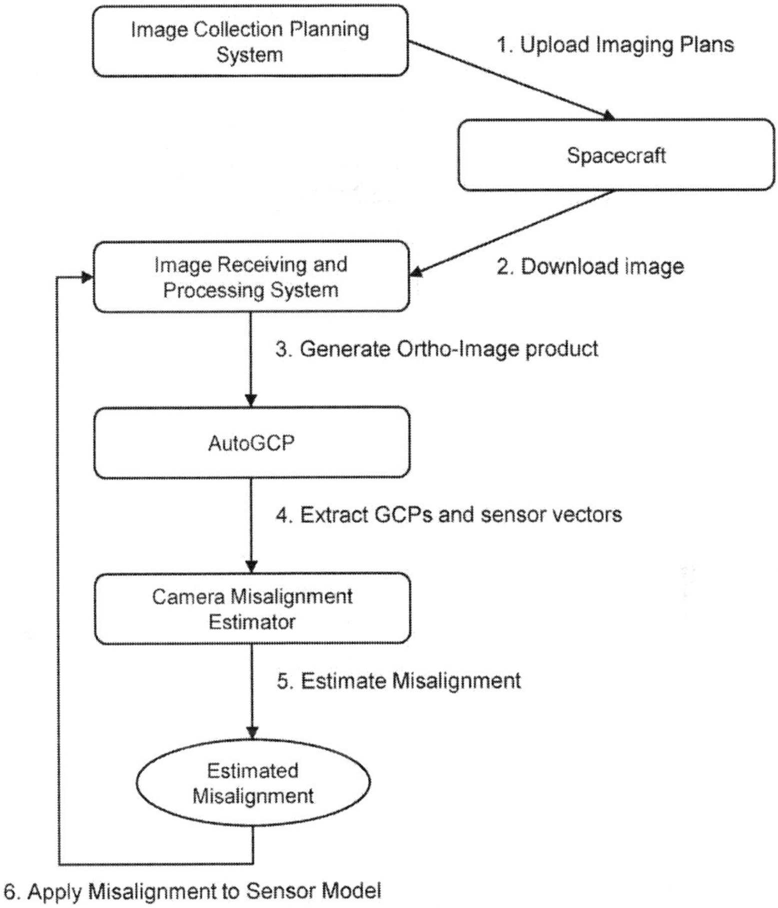

Figure 2. Structural process of camera misalignment estimation.

Attitude Sensors

Deimos-2 has two star-trackers for absolute attitude sensing, as well as four gyroscopes for relative attitude sensing. Figure 3 illustrates the exterior configuration of Deimos-2. The red cones are the field-of-view of star-trackers. The gyroscopes are internally equipped. Because the star-trackers are embedded at the opposite side to each other, one of or both star-trackers are selected for the absolute attitude sensing depending on the position and attitude of the satellite during imaging. The electro-optical camera, which is the main payload of Deimos-2, is at the opposite side of the solar panels.

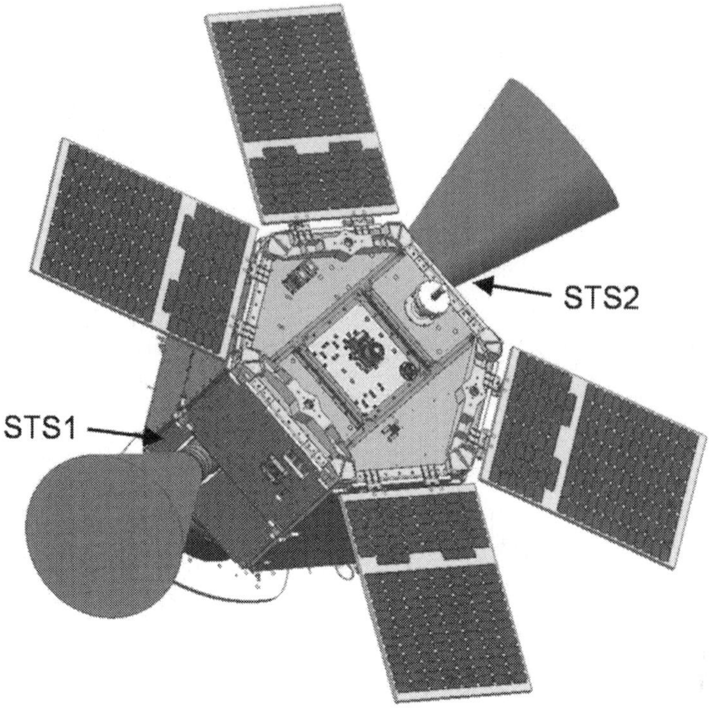

Figure 3. Mounting configuration of Deimos-2 star trackers.

Figure 4 shows the relationship of sensors/actuators and the optical cubes. The sensors and actuators for attitude determination and control are aligned by using optical cubes (OC) before launch. Optical cubes are used as a reference object for precision alignment in satellite manufacturing. The components that need inter-alignment such as imaging sensors and attitude sensors have their own optical cubes that are carefully aligned with them. The alignment between optical cubes are precisely measured using theodolites. Although pre-launch measurement was done for some components, the alignment needs to be calibrated after launch due to launch shock, outgassing, zero-gravity, thermal effect, *etc.* The alignment between the star-trackers and their optical cubes were unknown due to difficulty in ground measurement. The measurement of angles between optical cubes is used to build a rotation matrix that converts a vector of one component's frame to the vector of another component's frame. For Deimos-2, the attitude data in spacecraft ancillary which IRPS uses is based on Camera OC frame, which is also known as Attitude frame. The attitude measured using star-trackers is converted to Camera OC frame. The camera misalignment estimation that described in this paper estimates the discrepancy between this attitude and the GCPs acquired from the image that is in the unknown Camera frame. Hence, the combined error between the attitude sensor and the camera is estimated.

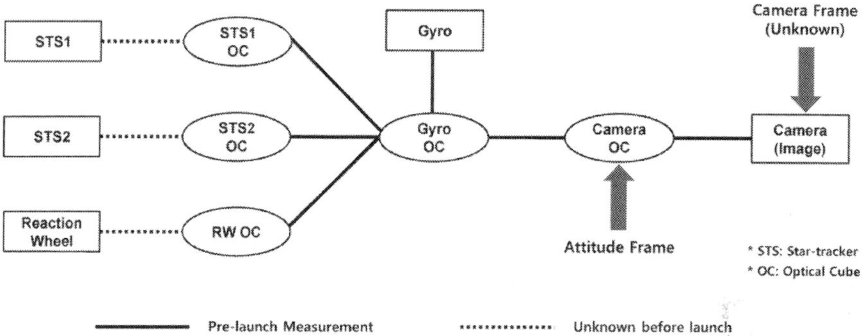

Figure 4. Alignment map of attitude sensors and camera.

The alignments between sensors/actuators and optical cubes are calibrated after launch by space-level calibration. Space-level calibration is a series of operations performed during early operation phase, which consists of attitude calibration maneuver and stellar imaging for alignment calibration.

Camera Geometry

Figure 5 shows the sensor geometry of Deimos-2 camera system. The alignment of the camera system is aligned with the spacecraft body using the Camera OC (Optical Cube) A. The alignment between the coordinates system of Camera OC A, CS_{cube}, and the coordinates system of the camera focal plane, CS_{det}, is measured before launch. Two focal plane assemblies (FPA1, FPA2) are aligned to have ground footprints perpendicular to the flight direction with a small overlap (100 panchromatic pixels). The linear TDI CCD arrays are positioned with the order of panchromatic, blue, green, red, and near-infrared in along-track direction. The 6115-th pixel in panchromatic FPA2 is the reference detector that defines the origin and Z-axis direction of the camera frame, which is CS_{det}.

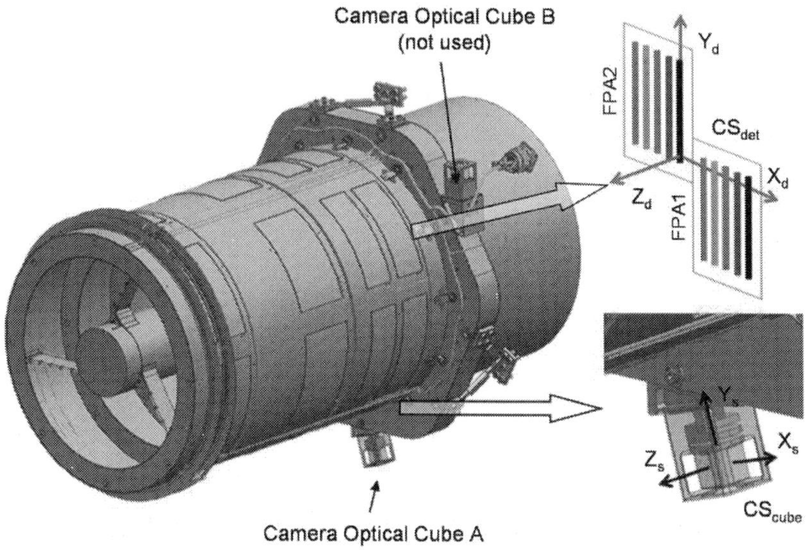

Figure 5. Geometry of Deimos-2 camera.

Reference Frames

In this section, the reference frames (a.k.a. reference coordinate systems) used in this paper are defined in detail. For well-known frames such as Earth-centered inertial (ECI) and earth-centered earth-fixed (ECEF), J2000 and WGS-84 are used respectively. The definitions of the Deimos-2 attitude frame and camera frame are presented in Table 1.

The attitude frame is the basis of attitude information in spacecraft ancillary data. It is defined by Camera OC in Deimos-2 as shown in Figure 4. Note that the definitions of the camera system and attitude reference system in Table 1 are the same for Deimos-2. Other spacecrafts may employ different definitions, requiring a rotation matrix to calculate the camera frame from a spacecraft attitude. It is represented by the rotation matrix $\mathrm{R}_{\mathrm{Cam}}^{\mathrm{Att}}$ and its inverse R_{Att}^{Cam} in this paper. They are identity matrices for Deimos-2.

Table 1. Definitions of Attitude Frame and Camera Frame.

	Attitude Frame	Camera Frame
Origin	Spacecraft center of mass	Reference detector
Z	Boresight vector of the reference detector	Boresight vector of the reference detector
Y	Longitudinal direction of linear CCD (along with Geometric Y axis direction)	Longitudinal direction of linear CCD (along with Geometric Y axis direction)
X	Y × Z (right hand rule)	Y × Z (right hand rule)

The camera frame is used by sensor vectors, which point at the corresponding detector cells. IRPS and AutoGCP use sensor vectors to convert image coordinates to geographic coordinates and *vice versa*.

Image Planning

The first step is planning imaging scenarios for camera misalignment estimation. It is important to cover all of the possible imaging scenarios that can occur during normal operation. The criteria that needs to be considered for a reliable and accurate estimation are as below.

- Tilt angle coverage
- Attitude sensor coverage
- Global coverage
- Terrain variation
- Visual distinctiveness

Tilt angle coverage: The image dataset for camera misalignment estimation needs to use various tilt angles that the spacecraft provides to get a reliable estimation result.

Attitude sensor coverage: If the spacecraft has multiple attitude sensors, those sensors are likely to have different misalignment toward the attitude frame. Since the attitude misalignment also introduces localization error like the camera misalignment, the estimation of the camera misalignment is influenced by the misalignment of attitude sensors. Therefore, the camera misalignment needs to be estimated separately for each sensor group. Deimos-2 is equipped with two star-trackers, and one or both of them are selected depending on the position and attitude of the spacecraft at the time of imaging.

Global coverage: Ground targets needs to be distributed all over the world covering all longitude and latitude ranges. It is not only beneficial for obtaining reliable results, but it can also show a possible relationship between the localization error and the location of a target.

Terrain variation: For the characteristics of ground targets, flat terrain is preferred in order to avoid possible discrepancy of digital elevation model (DEM) and real terrain.

Visual distinctiveness: Area with many visual features are also preferred. However, downtown areas with sky-scrappers are not recommended as tall buildings may add additional localization error, especially at a high tilt angle.

Automated GCP Extraction

The process of the camera misalignment estimation requires the comparison of the calculated localization with the true localization for GCPs in the images. However, it is exhausting work to obtain a large number of GCPs from many images manually.

Automated GCP extraction software generates GCPs from a geo-coded ortho-image and the corresponding reference dataset and measures the localization error between them. The reference dataset could be geo-coded ortho-rectified images such as LandSat-5 imagery or publicly available satellite imagery service such as Google Maps™. Figure 6 illustrates the process of the automated GCP extraction.

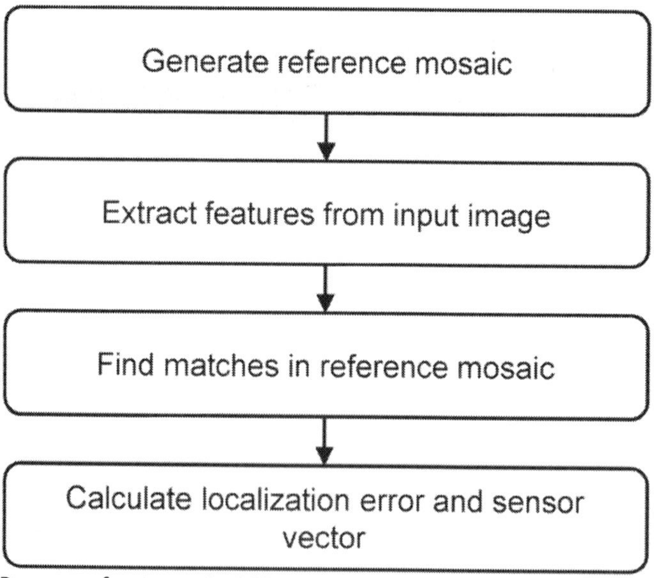

Figure 6. Process of automatic GCP extraction.

The first step is to generate a reference image mosaic. The search region is determined by the geo-location information embedded in the input image. The input image before the camera misalignment correction can be located at a different place from the actual location. In order to cover all possible areas, the search region is set to an expansion of the original input image boundary by a certain ratio. A reference mosaic image for the search region is then generated from the reference image database.

In order to extract image features from the input image, GFTT (Good Feature To Track) feature detector in OpenCV, which is based on Shi and Tomoshi's method [23], is used for corner detection. To ensure features are distributed all over the image, the minimum distance between features d is enforced as follows,

$$d = \alpha \left(\frac{w \cdot h}{N} \right)^{0.5} \tag{1}$$

where N is the maximum number of features to be extracted, w and h are image width and height, α is a flexibility parameter which is decided empirically to be 0.5.

The corresponding features in the reference mosaic are found by using a Template Matching method. A comparison region, which is as large as the input image boundary, is chosen from the search region. The features in the input image are searched from the comparison region. Finding the related features within a comparison region reduces the chance of false matches, in comparison to searching in the whole search region. Normalized cross-correlation metric is used as a similarity measure. Each feature of the input image is searched within the comparison region for the similar feature by template matching. The size of search window for template matching is large (e.g., 150 pixels), and the search offset for matching is (0, 0) at the beginning. They are optimized as the number of matches increases as follows,

$$\begin{gathered} e_i = u_i - v_i, i = 1 \ldots N \\ p = median(e_i) \\ s = \beta \times stdev(e_i - p) \end{gathered} \tag{2}$$

where the offset error ei is the difference between the input image feature vi and the comparison region feature ui. N is the total number of features. The search offset p is the median of offset errors of matches, eliminating outliers. The search window size s is the standard deviation of median-centered offset times margin factor β. In this paper, 3.0 is used for β.

When the searching for matches is completed, a homography matrix is estimated by using RANSAC method, so that the outliers are filtered out and inliers are reprojected. The comparison region is moved to another location within the search region to find better matches. The simulated annealing method was used to choose the next location of the comparison region. It chooses a random location at first, but it reduces its randomness as the iteration goes on, so that it can search near the last successful region. The iteration stops when the number of matches is the same as that of the last iteration. Figure 7 shows an example of feature matching results.

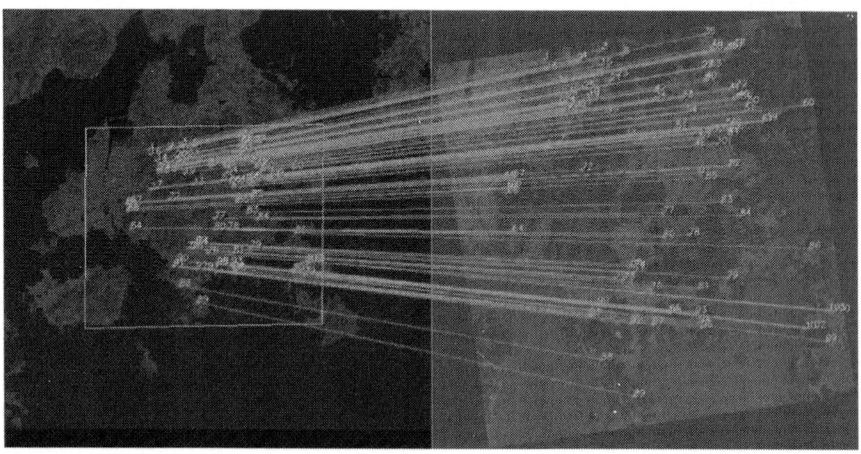

Figure 7. Correspondences between reference ortho-photo and input ortho-image.

Since the reference mosaic image is geo-referenced, the input image coordinates and the corresponding true geographic locations for each matched pair can be extracted, as well as the localization error of each pair. In order to use GCPs for camera misalignment estimation, the image coordinates need to be converted to sensor vectors.

A sensor vector is a unit direction vector in the camera frame pointing to the detector cell that corresponds with the image coordinates. In order to get a sensor vector from the image coordinates of the input image, the physical camera model that generated the input image performs the inverse mapping.

For each GCP, its IRPS-calculated geo-location, which differs from the true geo-location, could be re-projected onto the focal plane using the spacecraft position and attitude. Since, it is a push-broom sensor, it is not straightforward because the spacecraft takes an image over a certain period. The position and attitude of the moment that the GCP was imaged are necessary for re-projection, however, the time of that moment is unknown. In order to find the time t, an iterative method is used. Firstly, the center time of the imaging duration is used as the initial value of t, and the position and attitude of it is used for re-projection. The re-projected point on the focal plane shall not fall on the sensor because the position and attitude of the wrong time were used. The flight-direction distance between the sensor and the re-projected point is converted to time, and update t with it to move the point to the sensor. The ground point is re-projected with the position and attitude of the new time t. It requires a few iterations until the re-projected point falls on a detector cell. When the point is pointing a detector of the sensor, the vector pointing the re-projected point from the origin of the camera frame is the sensor vector of the GCP. Leprince *et al.* [24] presented a similar inverse mapping algorithm, however, our approach provide less time-complexity as it uses only one inverse mapping per iteration while theirs require two forward mappings per iteration to use the gradient minimization method.

The results of this process (*i.e.*, GCP geo-location, sensor vector, time t, localization error of GCPs) are stored for camera misalignment estimation.

Figure 8 and Figure 9 show the output of AutoGCP software, which performs the work described in this section. The localization error ε of each GCP is calculated as follows,

$$\varepsilon = x_{irps} - x_{ref} \tag{3}$$

where xirps is the UTM coordinate calculated by IRPS using the spacecraft ancillary, and xref is the UTM coordinate of the reference image.

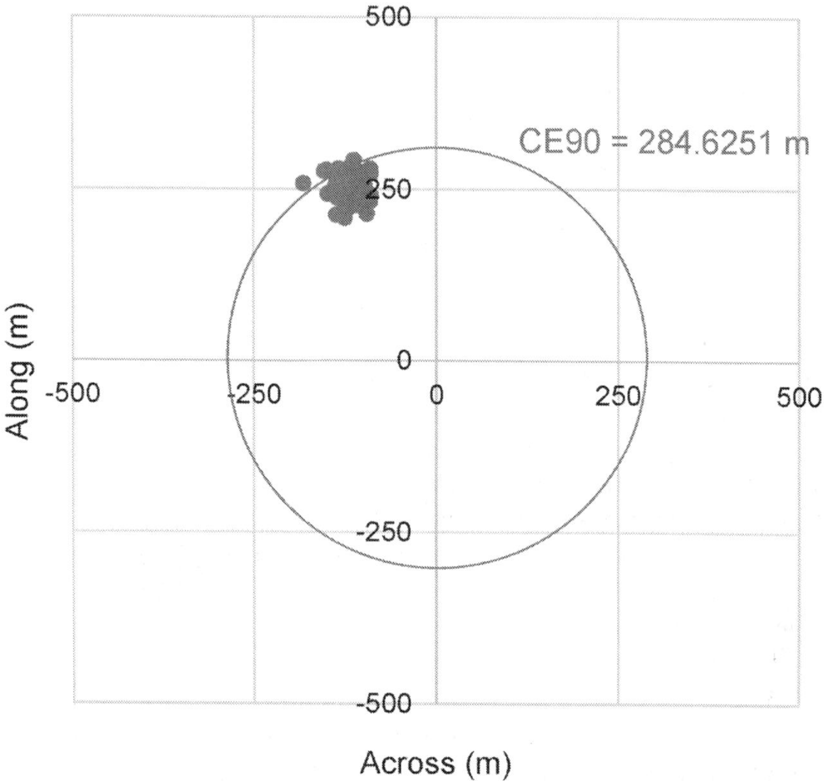

Figure 8. Measured localization error.

An important coordinates system for localization error analysis is spacecraft across/along-track coordinates system, which is defined by increasing column/line number direction. The analysis of localization errors in across/along-track direction makes it possible to analyze the errors in comparison with other imaging parameters such as tilt angles and target locations. Since the reference image and input ortho-image are generally in UTM projection, the localization error in UTM projection could be easily calculated. The conversion to spacecraft across/along-track direction from UTM X/Y direction could be done using the flight direction vector. Figure 10 shows the comparison of UTM X/Y direction and spacecraft across/along-track direction. Note that the along-track (+) direction is defined as the opposite of the flight direction in this paper, even though it is customary to set them to the same direction.

Figure 9. Extracted GCPs.

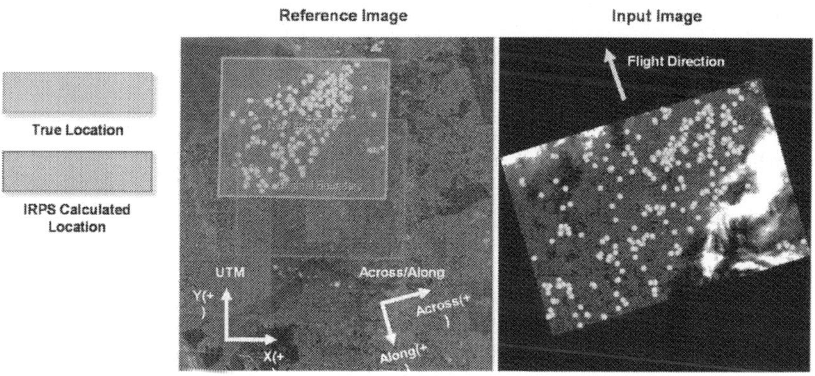

Figure 10. Comparison of UTM and across/long-track direction.

Error Pattern Analysis

Analyzing the pattern of localization error helps to find the existence of camera misalignment as well as other issues such as time drift. The estimation and correction of camera misalignment is meaningful when there is an obvious pattern of camera misalignment. It also is useful to know what type and amount of camera misalignment are there before estimation.

If a camera misalignment exists, a correlation between the tilt angle and localization error can be observed. Figure 11 shows the plots of localization error on various camera misalignments. Note that changing the sign of the misalignment will create localization errors of the opposite sign.

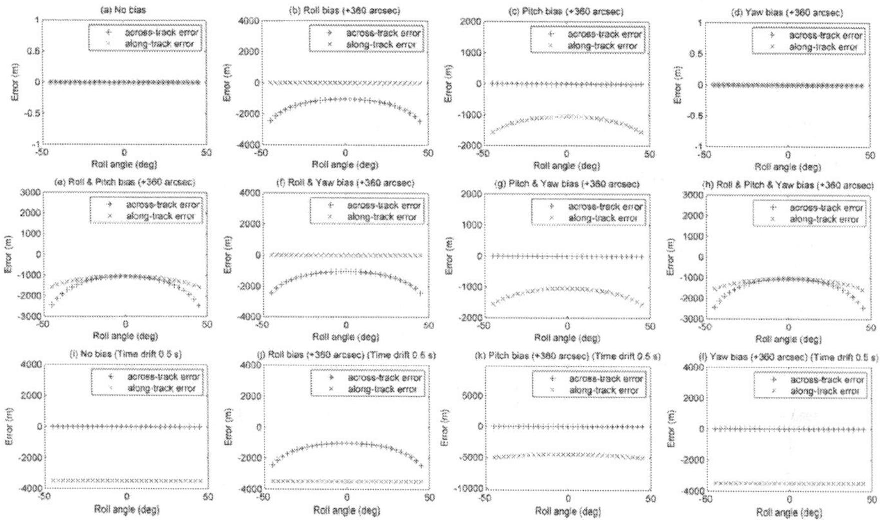

Figure 11. Localization errors by camera misalignment: (**a**) No bias, (**b**) Roll bias (+360 arcsec), (**c**) Pitch bias (+360 arcsec), (**d**) Yaw bias (+360 arcsec), (**e**) Roll & Pitch bias (+360 arcsec), (**f**) Roll & Yaw bias (+360 arcsec), (**g**) Pitch & Yaw bias (+360 arcsec), (**h**) Roll & Pitch & Yaw bias (+360 arcsec), (**i**) No bias (Time drift 0.5 s),(**j**) Roll bias (+360 arcsec) (Time drift 0.5 s), (**k**) Pitch bias (+360 arcsec) (Time drift 0.5 s), (**l**) Yaw bias (+360 arcsec) (Time drift 0.5 s).

Whilst roll and pitch biases display distinct patterns, yaw bias does not show a strong pattern in Figure 11d. It is because the yaw angle of the camera misalignment has small effect in localization error. The localization error ε that is generated by the yaw error at the end of the CCD array can be calculated as follows,

$$\varepsilon = \frac{1}{2}\vartheta_z LD \tag{4}$$

where ϑz is the yaw-bias in radian, L is the CCD array length, and D is the pixel GSD. Because of its relatively small error even with high tilt angles, it is difficult to estimate the yaw bias.

Sometimes, the ancillary data, which contains position and attitude, can have time offset or drift error. If time offset exists, along-track error has an offset as depicted in Figure 11i–l. In this case, it is difficult to relate to one of the biases in Figure 11b–h. In case of time drift, it should be corrected before estimating the camera misalignment. Since there are many sources that can cause time drift phenomena, the estimation of time drift is not discussed in this paper.

Another major source of localization error that can be confused with camera misalignment is attitude misalignment. The misalignment in the attitude frame affects the control and determination of the spacecraft attitude, resulting localization errors. Figure 12 shows the plots of localization error on various attitude misalignments. While the localization error induced by the yaw bias in the camera frame is hardly observable due to narrow field of view as shown in Figure 11d, it is clearly visible in Figure 12c that the yaw bias in the attitude frame generates large error as the tilt-angle increases.

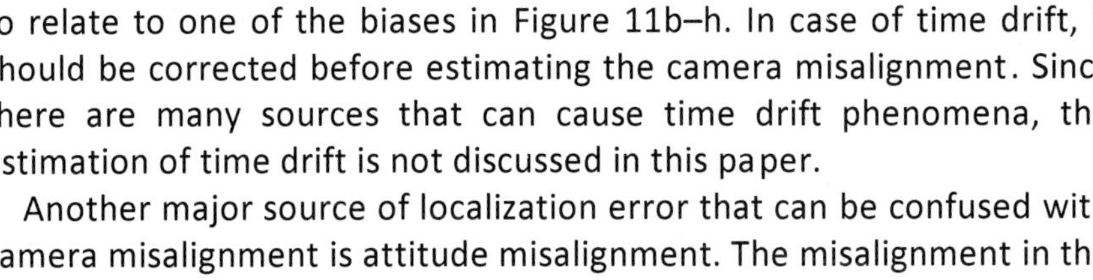

Figure 12. Localization errors by attitude misalignment: (**a**) Roll bias (+360 arcsec), (**b**) Pitch bias (+360 arcsec), (**c**) Yaw bias (+360 arcsec).

There are many other error sources of localization error such as star-trackers, gyroscopes, focal plane assemblies, sensor arrays, or somewhere between them. Only the sum of all misalignments from the attitude sensors to the camera image plane could be observed from image GCPs and attitude data. It is important to categorize datasets that can have different misalignment for effective analysis. In Section 3.2, Since Deimos-2 has two star-trackers and they are selected depending on the spacecraft attitude, its localization errors were analyzed for star-tracker selections.

In case that localization errors and tilt angles cannot be correlated, the cause of error may not be camera misalignment. They can be sensor malfunctioning, on-board software errors, or inconsistency of ancillary data. It is also recommended to analyze the relationship between the localization errors and target locations (longitude, latitude). Depending on the thermo-elastic characteristics of the spacecraft or the orbit design, the location of the target can affect the localization error.

Camera Misalignment Estimation

This section describes a mathematical model for camera misalignment estimation. The model describes the relationship between spacecraft position, attitude, sensor vector, and GCP. The key idea of the proposed estimation model is to calculate the boresight misalignment, which is the angular error between the camera frame and the true camera frame, using GCPs and spacecraft positions/attitudes. In order to simplify the problem, the origins of the attitude frame and the camera frame are assumed to be the same.

Assuming that there is no camera misalignment, the geometrical relationship can be modelled as follows:

$$p_S^{Cam} = R_{Att}^{Cam} R_{t_{ECI}}^{Att} p_G^{ECI} \tag{5}$$

where,

$$p_S^{Cam} = \left[\frac{T_S^{Cam}}{\|T_S^{Cam}\|}\right], T_S^{Cam} = \begin{bmatrix} x \\ y \\ -f \end{bmatrix}, p_G^{ECI} = \left[\frac{T_G^{ECI} - T_{t_S}^{ECI}}{\|T_G^{ECI} - T_{t_S}^{ECI}\|}\right] \tag{6}$$

p_S^{Cam} is a sensor vector, which is a unit vector of the vector T_S^{Cam}, which points the detector pixel on the focal plane at which a GCP is located. R_{Att}^{Cam} and $R_{t_{ECI}}^{Att}$ are rotation matrices from attitude frame to camera frame, and from ECI frame to attitude frame, respectively, at time t. p_G^{ECI} is an object vector in ECI frame pointing the GCP from the spacecraft. T_G^{ECI} is a position vector of ground object G in ECI frame. $T_{t_S}^{ECI}$ is a position vector of spacecraft in ECI frame at time t. Since the geo-locations of GCPs are generally in LLH or ECEF, it is important to convert the coordinate system to ECI to get p_G^{ECI}.

In order to take account for the camera misalignment, a bias-correction matrix $R_{Cam}^{TrueCam}$ is added:

$$p_S^{TrueCam} = R_{Cam}^{TrueCam} R_{Att}^{Cam} R_{t\,ECI}^{\ Att} p_G^{ECI}$$

(7)

In practice, the amount of misalignment is less than one degree (*i.e.*, 3600 arcsec). Therefore, an approximation for infinitesimal rotations can be used to solve Equation (7). Supposed that the misalignment is an Euler-angle rotation $\vartheta_{TrueCam}$,

$$\vartheta_{TrueCam} = \begin{bmatrix} \vartheta_x \\ \vartheta_y \\ \vartheta_z \end{bmatrix} \quad (8)$$

where ϑ_x, ϑ_y, ϑ_z are roll, pitch, yaw bias respectively, it can be approximated as follows [25,26]:

$$R_{Cam}^{TrueCam} \approx I + [\vartheta_{TrueCam} \times] \quad (9)$$

$$[\vartheta_{TrueCam} \times] = \begin{bmatrix} 0 & -\vartheta_z & \vartheta_y \\ \vartheta_z & 0 & -\vartheta_x \\ -\vartheta_y & \vartheta_x & 0 \end{bmatrix}$$

(10)

The error of the small angle approximation of sine and cosine function is less than 1% for angles smaller than 14 degrees (≈50,000 arcsec).

In consequence, the equation for a sensor vector and a unit object vector in the Equation (7) can be expanded as below.

$$p_S^{TrueCam} \approx (I + [\vartheta_{TrueCam} \times]) R_{ECI}^{Cam} p_G^{ECI}$$

(11)

$$= R_{ECI}^{Cam} p_G^{ECI} + [\vartheta_{TrueCam} \times] R_{ECI}^{Cam} p_G^{ECI}$$

(12)

$$= R_{ECI}^{Cam} p_G^{ECI} - [R_{ECI}^{Cam} p_G^{ECI} \times] \vartheta_{TrueCam}$$

(13)

Equation (13) can be rewritten as follows, which forms a normal equation for linear least squares (Ax = b):

$$[R_{ECI}^{Cam} p_G^{ECI} \times] \vartheta_{TrueCam} = R_{ECI}^{Cam} p_G^{ECI} - p_S^{TrueCam} \quad (\mathrm{Ax} = \mathrm{b})$$

(14)

In addition to the Equation (14), a weight matrix W is introduced to adjust weight of each observation and to force pre-defined bias to certain axes as below:

$$\mathrm{WAx} = \mathrm{Wb} \tag{15}$$

where the weight matrix W is defined as:

$$W_{(3(N+1)\times 3(N+1))} = \begin{bmatrix} W_1 & & & \\ & \ddots & & \\ & & W_N & \\ & & & r \end{bmatrix},$$

$$W_{i=1...N} = diag\left(\frac{1}{\sigma_i^2} \; \frac{1}{\sigma_i^2} \; \frac{1}{\sigma_i^2}\right), r_{(3\times3)} = \begin{bmatrix} \frac{1}{r_x^2} & & \\ & \frac{1}{r_y^2} & \\ & & \frac{1}{r_z^2} \end{bmatrix} \tag{16}$$

Other elements (*i.e.*, A, b, x) in the Equation (15) are also extended as follows,

$$W_{i=1...N} = diag\left(\frac{1}{\sigma_i^2} \; \frac{1}{\sigma_i^2} \; \frac{1}{\sigma_i^2}\right), r_{(3\times3)} = \begin{bmatrix} \frac{1}{r_x^2} & & \\ & \frac{1}{r_y^2} & \\ & & \frac{1}{r_z^2} \end{bmatrix}$$

(17)

$W_{i=1...N}$ sets the weight for each observation in accordance with the accuracy of observation. In case that every observation has the same accuracy, the weight is set to 1.0. However, it can also be set to a certain weight calculated from the accuracy of observation, such as the geo-accuracy of a GCP, sensor vector, and position/attitude knowledge.

V and r could be used to set the bias of some axes to given values. r is a parameter for choosing whether the pre-assigned bias will be enforced or not. Setting the element of r to a very small number (e.g., $10^{-1°}$) forces the assigned angle in V to be the bias of the given axis, whereas setting it to a very large number (e.g., $10^{1°}$) means V is ignored. It is useful when the visibility of a certain axis is expected to be very low due to narrow field of view; for instance, the yaw axis of high-resolution push-broom imaging sensors.

Solving Equation (15) gives a least squares solution of $\vartheta_{TrueCam}$, which contains estimated camera misalignment angles in radian. In order to compensate the camera misalignment during ortho-image generation, the alignment of camera frame from attitude frame needs to be updated using $\vartheta_{TrueCam}$. The new rotation matrix R'^{Cam}_{Att} is calculated as follows,

$$R'^{Cam}_{Att} = DCM(\vartheta_{TrueCam})R^{Cam}_{Att} \quad (18)$$

where DCM($\vartheta_{TrueCam}$) is a direction cosine matrix for $\vartheta_{TrueCam}$, and R^{Cam}_{Att} is the original rotation matrix. The order of rotations is irrelevant for infinitesimal angles [25,26].

EXPERIMENTAL SECTION

Simulation

Data

The objectives of simulation tests are to prove the validity of the estimation model and obtain expected accuracy before the application to real image data. The test environment was set up as illustrated in Figure 13. Three additional components were used to simulate an actual spacecraft, compared to Figure 2.

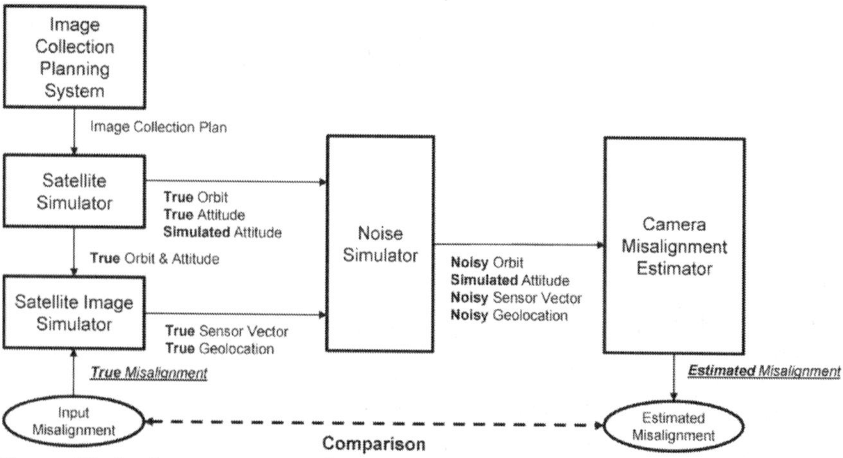

Figure 13. Environment setup for simulation test.

Satellite Simulator simulated the spacecraft system in space environment. It ran the same on-board software emulating all the position/attitude sensors and actuators of Deimos-2. As it complied the design specification of Deimos-2 hardware, it could simulate realistic behavior of a spacecraft on orbit. Image Simulator generated simulated imagery taken by the simulated spacecraft. It also generated sensor vectors and their corresponding geo-locations, which were the output of AutoGCP. Noise Simulator added noise to orbit data, sensor vectors, and geo-locations for realistic simulation.

Table 2 shows the categories of the experiments. The first category is the *no noise* category, which verifies if the proposed estimation model is capable of estimating camera misalignment in ideal cases. The second category is the *noise* category, which provides the estimated accuracy for real cases.

Table 2. Image Simulation Setting.

	No Noise			Noise		
Number of images		1			Same	
Size of image		6000 × 6000			Same	
Attitude (deg.)	Roll	Pitch	Yaw		Same	
	−4.243	+0.109	0			
Number of GCPs per image		270			Same	
Spacing between GCPs		350 m			Same	
Position Error (Gaussian) (meter)		0 m		Along	Across	Radial
				30	5	5
Attitude Knowledge (3σ) (arcsec)		0		Roll	Pitch	Yaw
				360	360	360
Planimetric GCP Error (Gaussian)		0 m			30 m	

Image Simulator is capable of setting alignment biases to simulate camera misalignment. In the experiments, eight cases with different combinations of bias values were tested. Camera misalignment Estimator estimated the bias for each case. The assigned bias values and estimated bias values were compared.

Results

Table 3 shows the results of the *no noise* category. The error of estimation model was up to 0.054 arcsec for all test cases. As 1 arcsec error corresponds to approximately 3.3 m on ground, it is proven that the model provides sub-meter accuracy for ideal cases. The error of 0.054 arcsec residual under zero noise seems like numerical error, since 0.01 arcsec is 2.61799398e^{-7} radian.

Table 3. Misalignment Estimation Results of *No Noise* Case.

True Bias (arcsec)			Estimated Bias (arcsec)			Error (arcsec)		
Roll	Pitch	Yaw	Roll	Pitch	Yaw	Roll	Pitch	Yaw
0	0	0	−0.001	0.006	−0.001	0.001	−0.006	0.001
100	0	0	99.999	0.006	−0.001	0.001	−0.006	0.001
0	100	0	−0.001	100.006	−0.001	0.001	−0.006	0.001
0	0	100	−0.001	0.006	99.999	0.001	−0.006	0.001
100	0	100	99.999	0.054	99.999	0.001	−0.054	0.001
0	100	100	−0.049	100.006	99.999	0.049	−0.006	0.001
100	100	0	99.999	100.006	−0.049	0.001	−0.006	0.049
100	100	100	99.951	100.054	99.951	0.049	−0.054	0.049

Table 4 shows the results of *noise* category. The errors were around 30 arcsec in roll and pitch biases and around 400 arcsec in yaw bias. In comparison with the sensor accuracy in Table 2, which is 360–432 arcsec, the accuracy of the roll/pitch biases estimation is better than the hardware specification. Given that the attitude error is mainly made of bias and random error, the influence of sensor's random error in camera misalignment estimation is reduced as more and more data samples are used. However, the yaw bias might still be difficult to estimate accurately, because of the narrow field-of-view of a high-resolution push-broom type sensor. It is not problematic in practice, because it also means that the influence of the yaw bias in the camera frame is relatively low.

Table 4. Misalignment Estimation Results of *Noise* Case.

True Bias (arcsec)			Estimated Bias (arcsec)			Error (arcsec)		
Roll	Pitch	Yaw	Roll	Pitch	Yaw	Roll	Pitch	Yaw
0	0	0	−13.466	13.466	23.545	−23.545	0.659	−0.659
100	0	0	87.880	12.120	29.495	−29.495	−397.391	397.391
0	100	0	−12.542	12.542	129.351	−29.351	259.387	−259.387
0	0	100	−14.160	14.160	20.218	−20.218	132.343	−32.343
100	0	100	86.432	13.568	22.535	−22.535	−140.799	240.799
0	100	100	−14.090	14.090	119.881	−19.881	152.769	−52.769
100	100	0	86.035	13.965	120.356	−20.356	−129.372	129.372
100	100	100	86.431	13.569	123.715	−23.715	190.704	−90.704

In case that the observability of yaw bias is expected to be low, the estimation model could be set to fixate the yaw bias to zero. Table 5 shows there was no noticeable change in the accuracy of roll and pitch bias when the yaw bias was not estimated.

Table 5. Misalignment Estimation Results of *Noise* Case with Yaw Bias = 0.

True Bias (arcsec)			Estimated Bias (arcsec)			Error (arcsec)		
Roll	Pitch	Yaw	Roll	Pitch	Yaw	Roll	Pitch	Yaw
0	0	0	−12.816	12.816	26.028	−26.028	0.000	0.000
100	0	0	86.396	13.604	22.465	−22.465	0.000	0.000
0	100	0	−14.029	14.029	120.620	−20.620	0.000	0.000
0	0	100	−11.551	11.551	32.314	−32.314	0.000	100.000
100	0	100	87.897	12.103	30.500	−30.500	0.000	100.000
0	100	100	−13.540	13.540	123.005	−23.005	0.000	100.000
100	100	0	86.711	13.289	123.092	−23.092	0.000	0.000
100	100	100	87.545	12.455	128.657	−28.657	0.000	100.000

In conclusion, the estimation model is capable of estimating the camera misalignment to sub-meter accuracy under ideal conditions. With realistic simulation data, the estimation accuracy is better than the accuracy of hardware while the yaw estimation does not affect the overall accuracy much.

Real Image

Data

The estimation framework presented in this paper was applied to Deimos-2 to measure and correct its camera misalignment. It began with listing up candidate targets in the world. Figure 14 shows the distribution of the ground target images used in the experiments.

The targets were focused on populated area containing rich features to obtain enough number of GCPs. The colors of markers (red, green, blue) in the figure represent which star-tracker was selected for on-board attitude determination. Deimos-2 has three star-tracker selection modes (STS 1, STS 2, STS Both) and the figure shows that all selections are included for the test images. Note that the selection is based generally on the location of the ground target as shown in the figure as well as the off-nadir tilt angle.

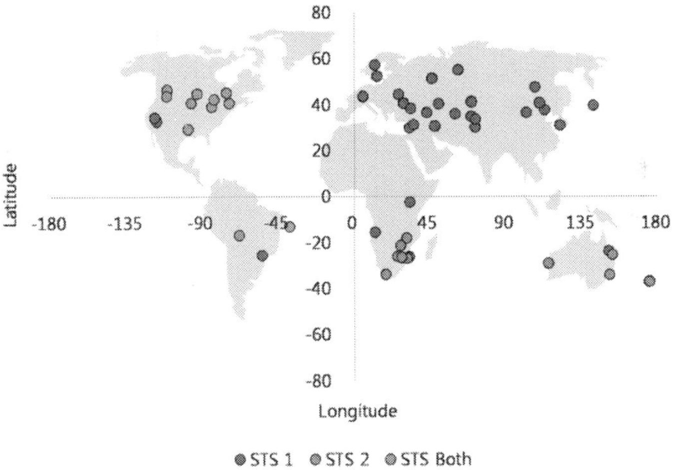

Figure 14. Ground target distribution of Deimos-2.

In order to investigate the difference between imaging modes, the images were taken in two imaging modes; Strip and Fast Multi-strip (FMS). In Strip mode, the spacecraft returns to house-keeping mode after imaging, whereas in FMS mode, it performs a maneuver and takes successive images before returning to house-keeping mode. Table 6 shows the number of images collected for camera misalignment estimation.

Table 6. Number of Images Collected For Deimos-2 Calibration.

Imaging Mode	# Images
Strip	37
FMS	51
Total	88

Figure 15a shows the distribution of imaging dates. The images were taken over a two-week period. There are gaps at 10 and 14 July 2014 because the spacecraft was dedicated to other priority tasks.

Figure 15b shows the distribution of roll-tilt angles. The images cover all of the roll angle range that Deimos-2 is capable. The pitch angle and yaw angle were all set to zero, since Deimos-2 does not provide pitch and yaw tilt in Strip and FMS imaging mode.

The images were processed to ortho-image products after download. The SRTM DEM which has 90 m accuracy was used for geometric rectification. AutoGCP software extracted GCPs and the corresponding sensor vectors from the ortho-images, and calculated the localization errors. The spatial resolution of reference mosaic for GCP extraction was set to 15 m, which corresponds to the geo-location accuracy of resulting GCPs, accounting for the accuracy of Deimos-2 attitude sensors.

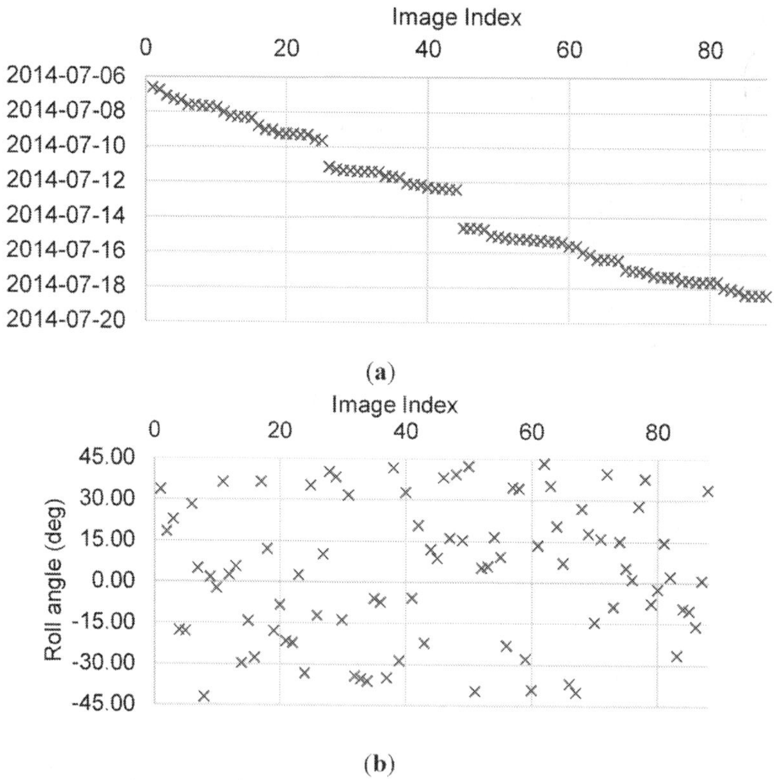

Figure 15. Distribution of (**a**) imaging dates and (**b**) roll-tilt angles.

In order to inspect the localization error, the mean μerror and standard deviation σerror of across-track and along-track errors are defined as follows,

$$\mu_{error} = \frac{1}{m}\sum_{i=0}^{m} e_i$$

$$\sigma_{error} = \sqrt{\frac{1}{m}\sum_{i=0}^{m}(e_i - \mu_{error})^2}$$

$$e_i = \begin{bmatrix} \text{mean across-track error of GCPs in the i-th image} \\ \text{mean along-track error of GCPs in the i-th image} \end{bmatrix}$$

(19)

where m is the number of images, ei is the mean across-track and mean along-track error of i-*th* image. The mean error shows how much the localization errors of the images are biased, while the standard deviation represents the variance of localization errors of different images.

The localization error is presented in Table 7. For both Strip and FMS modes, the mean error was significantly larger than the standard deviation of error. As camera misalignment causes biased error, it is assumable that there was a camera misalignment. There was no significant difference for the mean errors between imaging modes, meaning that it was unnecessary to estimate the camera misalignment for each imaging mode separately.

Table 7. Localization Error of Deimos-2 before Camera Misalignment Update.

		Across-Track Error (m)	Along-Track Error (m)	CE90 (m)
Strip	Mean	−158.73	235.12	402.10
	Std. Dev.	76.25	81.04	
FMS	Mean	−134.56	224.86	405.97
	Std. Dev.	111.53	91.86	
All	Mean	−144.72	229.18	405.97
	Std. Dev.	98.13	87.81	

Figure 16 shows the localization errors for different star-tracker selections. The X-axis is roll-tilt angles and the Y-axis is mean across-track error and mean along-track error. The observation of the localization errors between star-trackers separately led to a conclusion that the error pattern showed meaningful differences depending on the star-tracker selection. Therefore, it was necessary to estimate the camera misalignment for each star-tracker selection mode separately. The plots in Figure 16 also displays the patterns of roll bias and pitch bias shown in Figure 11.

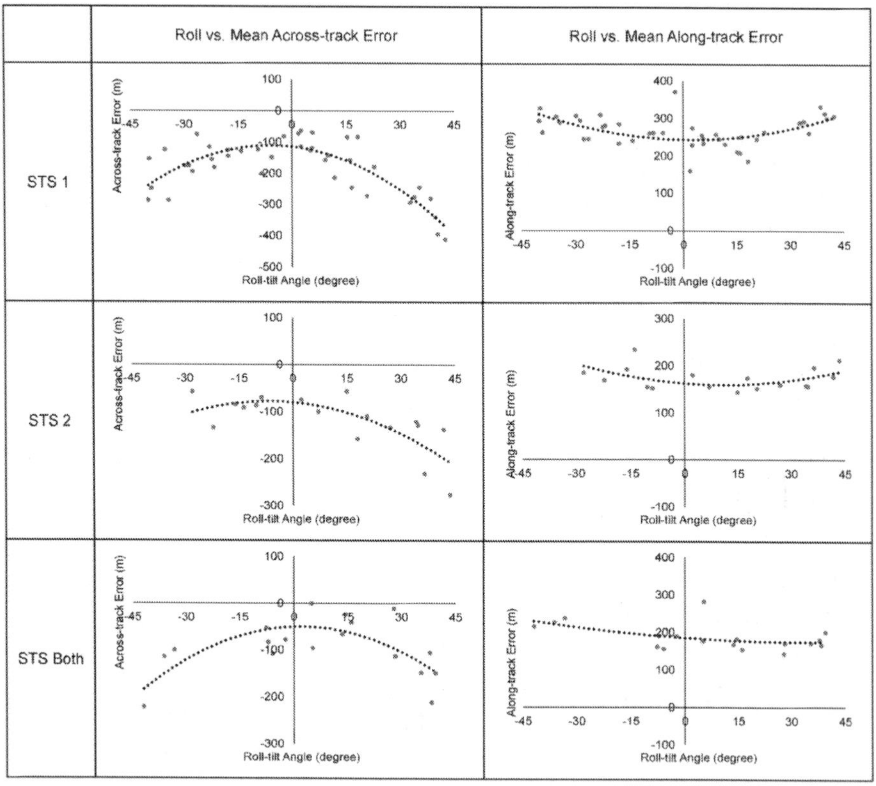

Figure 16. Localization error of Deimos-2 before camera misalignment estimation.

Among 88 images, 41 unsuitable images were eliminated by the following criteria.

- Position/attitude/imaging sensors were not properly working
- Number of GCPs is less than 200
- GCPs are located partially in the image
- Localization error does not follows the trend of other images
- Standard deviation of localization error within the image is high

Since Deimos-2 was in the initial commissioning phase, the sensors had some functionality issues. Twenty-three images out of 88 images were dropped due to sensor malfunctioning. The standard deviation of the localization errors of GCPs which belong to one image contains various meanings. A higher standard deviation is, however, largely due to the high elevation variation within the imaged area. Although DEM was used for ortho-image generation, the inaccuracy of DEM degrades the estimation accuracy. In this experiment, the SRTM DEM that has 90 m special resolution was used. Table 8 shows the number of images and GCPs which were used for the camera misalignment estimation for each star-tracker selection mode.

Table 8. Number of images and GCPs for camera misalignment estimation.

	# Images	# GCPs
STS 1	17	25,137
STS 2	18	18,956
STS Both	12	15,124

The camera misalignment was estimated for each star-tracker selection mode by using the corresponding GCPs and the satellite position and attitude data as described in Section 2.6. Because of the narrow field-of-view and limited accuracy of attitude knowledge of Deimos-2, the yaw bias estimation was excluded by setting the estimator to fixate the yaw misalignment to zero to avoid overfitting. The estimated misalignment was applied to the ground image processing software to regenerate all 88 ortho-image products by using the updated camera model.

3.2.2. Results

Table 9 shows the results of the camera misalignment estimation. The residual RMSE is about 1/5 of the original RMSE, showing that the localization errors could be well corrected by the estimated camera misalignment.

Table 9. Estimated Camera Misalignment of Deimos-2.

	Estimated Misalignment		RMSE	
	Roll (arcsec)	Pitch (arcsec)	Original (arcsec)	Residual (arcsec)
STS 1	47.93	−78.85	93.34	14.07
STS 2	27.98	−49.72	58.29	11.93
STS Both	22.97	−52.43	58.05	9.61

The localization errors were recalculated using the reprocessed ortho-images in order to compare them to the original errors. Table 10 shows the improvement of the localization accuracy after applying the estimated camera misalignment. "All" rows in the STS Selection column shows the statistics calculated using all images; STS 1, STS 2, and STS Both.

Table 10. Changes of Average Localization Error.

	STS Selection	Across-Track Error		Along-Track Error	
		Mean (m)	Std. Dev. (m)	Mean (m)	Std. Dev. (m)
Original	STS 1	−174.32	97.64	263.38	84.12
	STS 2	−121.51	54.97	188.34	66.91
	STS Both	−95.59	59.22	187.91	33.52
	All	−144.72	87.81	229.18	80.65
After Update	STS 1	13.66	73.92	-3.44	76.28
	STS 2	−10.69	42.02	22.21	69.66
	STS Both	−4.06	52.05	16.35	68.15
	All	4.02	63.53	6.95	73.18

Figure 17 illustrates the localization error of each image before and after applying the estimated camera misalignment, where X-axis is image index and Y-axis is across/along-track error. It can be seen that the bias of the localization errors was eliminated effectively by applying the estimated camera misalignment.

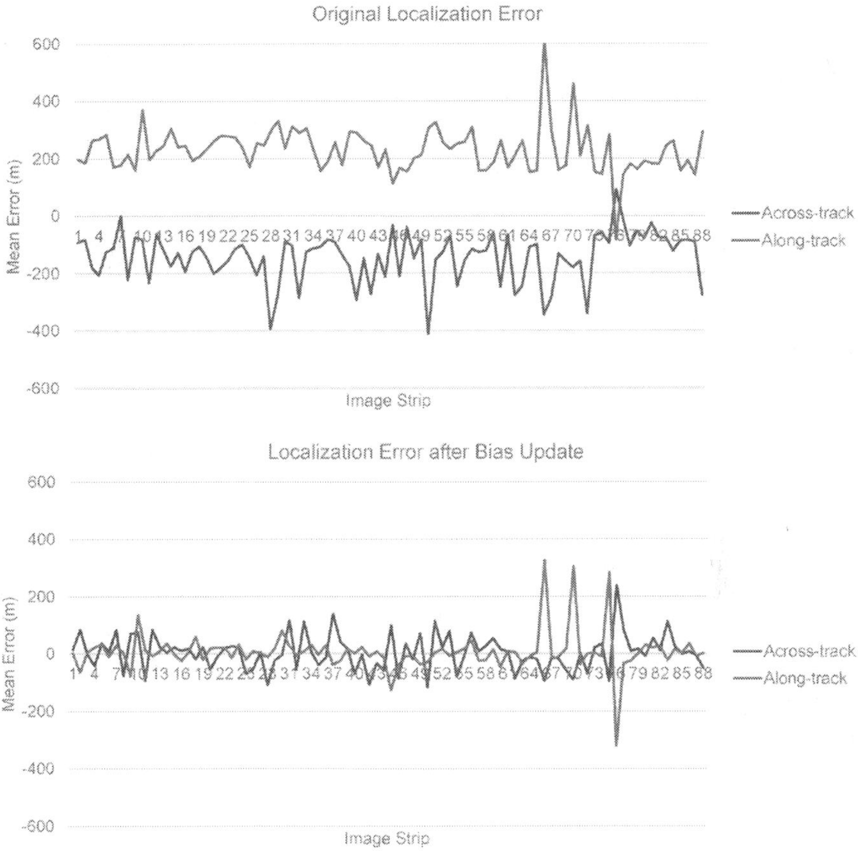

Figure 17. Changes of localization error per image.

Table 11 summarizes the results of the proposed technique including the CE90 geo-location accuracy achieved by using the proposed framework. Considering that the geo-location accuracy of GCPs was 15 m, the bias of images was well compensated to the GCP's accuracy level. The residual localization error for each image is due to the inaccuracy of spacecraft position/attitude knowledge and the inaccuracy of DEM. Deimos-2 is a cost-effective high-resolution spacecraft that equips attitude sensors that are less accurate than those of IKONOS, WorldView-1, or Pleiades-1A/1B. It limits the accuracy of camera misalignment estimation as well as the localization accuracy after misalignment update.

Table 11. Changes of Average Localization Error of All Images.

		# Images	Across-Track Error (m)	Along-Track Error (m)	CE90 (m)
Original	Strip	37	−158.73	235.12	402.10
	FMS	51	−134.56	224.86	405.97
	All	88	−144.72	229.18	405.97
After Update	Strip	37	1.16	0.86	116.09
	FMS	51	6.11	11.38	119.33
	All	88	4.02	6.95	119.33

CONCLUSIONS

This paper presented a framework for camera misalignment estimation and supporting results. The proposed framework consists of four steps: image planning, automated GCP extraction, error pattern analysis, and camera misalignment estimation.

A set of simulation experiments was performed to verify the validity of the proposed mathematical model. A satellite simulator and an image simulator were used for the generation of realistic simulation data. Various camera misalignment settings were simulated and compared with the estimation results. In this simulation experiments, it was proven that the proposed estimation model can achieve a sub-meter geo-location accuracy in ideal cases and respond to the characteristics of sensor noise specifications very well.

The application of the entire framework was demonstrated by using several tens of real image data covering all operational cases of the satellite in order to prove the feasibility of the proposed framework. The imaging planning decisions and the pattern analysis schemes in the proposed framework provided a guideline for the camera misalignment estimation activity during the initial commissioning period of a high-resolution satellite program. Hundreds of GCPs per image were extracted automatically for camera misalignment estimation. The estimated misalignment was applied to the ground image processor so that the localization error was reduced from 405.97 to 119.33 m in CE90.

The proposed framework is an effective and efficient method to estimate an accurate camera misalignment during a post-launch calibration and validation phase. As it provides precise and reliable measurements and estimation results, it may also reduce the efforts to pre-launch calibration activities. Although this paper focused on the camera misalignment estimation, the presented ideas can be applied to a wide range of applications. The presented GCP extraction algorithm provides many GCPs automatically which can be used for the calibration of other parameters such as focal length, line rate, and image distortion. It can also be attached to the image processing chain to improve the geo-location accuracy of images through a systematic geo-correction using GCPs. The analysis of localization error patterns is useful to discover many other errors such as telemetry error, time drift, gyro scale factor error, and other hardware or software errors of the satellite.

ACKNOWLEDGMENT

The authors would like to thank Deimos Castilla La Mancha for valuable supports including, but not limited to, Deimos-2 satellite planning and operation, image data reception and provision, and technical consultancy for this work.

AUTHOR CONTRIBUTIONS

The first author implemented the software, performed the experiments, and wrote the paper. The second author derived the camera misalignment estimation model and gave helpful comments and guidance to the first author.

CONFLICTS OF INTEREST

The authors declare no conflict of interest.

REFERENCES

1. Müller, R.; Lehner, M.; Reinartz, P. Evaluation of spaceborne and airborne line scanner images using a generic Ortho image processor. In Proceedings of High Resolution Earth Imaging for Geospatial Information, Hannover, Germany, 17–20 May 2005.
2. Tang, X.; Xie, J. Overview of the key technologies for high-resolution satellite mapping. *Int. J. Digit. Earth* **2012**, *5*, 228–240.

3. Chen, T.; Shibasaki, R.; Lin, Z. A rigorous laboratory calibration method for interior orientation of an airborne linear push-broom camera. *Photogramm. Eng. Remote Sens.* **2007**, *73*, 369–374.
4. Burner, A.W.; Snow, W.L.; Shortis, M.R.; Goad, W.K. Laboratory calibration and characterization of video cameras. *Proc. SPIE* **1990**, *1395*, 664–671.
5. Liu, T.; Burner, A.W.; Jones, T.W.; Barrows, D.A. Photogrammetric techniques for aerospace applications. *Prog. Aerosp. Sci.* **2012**, *54*, 1–58.
6. Radhadevi, P.V.; Solanki, S.S. In-flight geometric calibration of different cameras of IRS-P6 using a physical sensor model. *Photogramm. Rec.* **2008**, *23*, 69–89.
7. Bouillon, A.; Bernard, M.; Gigord, P.; Orsoni, A.; Rudowski, V.; Baudoin, A. SPOT 5 HRS geometric performances: Using block adjustment as a key issue to improve quality of DEM generation. *ISPRS J. Photogramm. Remote Sens.* **2006**, *60*, 134–146.
8. Breton, E.; Bouillon, A.; Gachet, R.; Delussy, F. Pre-flight and in-flight geometric calibration of SPOT5 HRG and HRS images. *ISPRS Int. Arch. Photogramm. Remote Sens. Spat. Inf. Sci.* **2002**,*34*, 20–25.
9. Dial, G.; Bowen, H.; Gerlach, F.; Grodecki, J.; Oleszczuk, R. IKONOS satellite, imagery, and products. *Remote Sens. Environ.* **2003**, *88*, 23–36.
10. Zhang, C.; Fraser, C.S.; Liu, S. Interior orientation error modelling and correction for precise georeferencing of satellite imagery. *Int. Arch. Photogramm. Remote Sens. Spat. Inf. Sci.* **2012**,*39B1*, 285–290.
11. Mulawa, D. On-orbit geometric calibration of the OrbView-3 high resolution imaging satellite. *Int. Arch. Photogramm. Remote Sens. Spat. Inf. Sci.* **2004**, *35*, 1–6.
12. Leprince, S.; Musé, P.; Avouac, J.-P. In-flight CCD distortion calibration for pushbroom satellites based on subpixel correlation. *IEEE Trans. Geosci. Remote Sens.* **2008**, *46*, 2675–2683.
13. Moorthi, S.M.; Kayal, R.; Krishnan, R.R.; Srivastava, P.K. RESOURCESAT-1 LISS-4 MX bands onground co-registration by in-flight calibration and attitude refinement. *Int. J. Appl. Earth Obs. Geoinf.* **2008**, *10*, 140–146.
14. Robertson, B.; Beckett, K.; Rampersad, C.; Putih, R. Quantitative geometric calibration & validation of the RapidEye constellation. In Proceeding of the 2009 IEEE International Geoscience and Remote Sensing Symposium, Cape Town, South Africa, 12–17 July 2009.
15. Takaku, J.; Tadono, T. PRISM on-orbit geometric calibration and DSM performance. *IEEE Trans. Geosci. Remote Sens.* **2009**, *47*, 4060–4073.
16. Delevit, J.M.; Greslou, D.; Amberg, V.; Dechoz, C.; de Lussy, F.; Lebegue, L.; Latry, C.; Artigues, S.; Bernard, L. Attitude assessment using Pleiades-HR capabilities. *Int. Arch. Photogramm. Remote Sens. Spat. Inf. Sci.* **2012**, *39B1*, 525–530.
17. Greslou, D.; de Lussy, F.; Delvit, J.M.; Dechoz, C.; Amberg, V. Pleiades-Hr Innovative Techniques for Geometric Image Quality Commissioning. *Int. Arch. Photogramm. Remote Sens. Spat. Inf. Sci.* **2012**, *39B1*, 543–547.

18. De Lussy, F.; Greslou, D.; Dechoz, C.; Amberg, V.; Delvit, J.M.; Lebegue, L.; Blanchet, G.; Fourest, S. Pleiades HR in flight geometrical calibration: Location and mapping of the focal plane.*Int. Arch. Photogramm. Remote Sens. Spat. Inf. Sci.* **2012**, *39*, 519–523.
19. Wang, M.; Yang, B.; Hu, F.; Zang, X. On-Orbit geometric calibration model and its applications for high-resolution optical satellite imagery. *Remote Sens.* **2014**, *6*, 4391–4408.
20. Jiang, Y.; Zhang, G.; Tang, X.; Li, D.; Huang, W. Detection and correction of relative attitude errors for ZY1-02C. *IEEE Trans. Geosci. Remote Sens.* **2014**, *52*, 7674–7683.
21. Müller, R.; Krauß, T.; Schneider, M.; Reinartz, P. Automated georeferencing of optical satellite data with integrated sensor model improvement. *Photogramm. Eng. Remote Sens.* **2012**, *78*, 61–74.
22. Klančar, G.; Blažič, S.; Matko, D.; Mušič, G. Image-based attitude control of a remote sensing satellite. *J. Intell. Robot. Syst.* **2011**, *66*, 343–357.
23. Shi, J.; Tomasi, C. Good features to track. In Proceedings of the IEEE Conference on Computer Vision and Pattern Recognition, Seattle, WA, USA, 21–23 June 1994.
24. Leprince, S.; Barbot, S.; Avouac, J.-P. Automatic and precise orthorectification, coregistration, and subpixel correlation of satellite images, application to ground deformation measurements.*IEEE Trans. Geosci. Remote Sens.* **2007**, *45*, 1529–1558.
25. Shuster, M.D. Survey of attitude representations. *J. Astronaut. Sci.* **1993**, *41*, 439–517.
26. Infinitesimal Rotation. Available online: http://rotations.berkeley.edu/?page_id=1682 (accessed on 15 September 2014).

CITATION

Seungwoo Lee and Dongseok Shin, On-Orbit Camera Misalignment Estimation Framework and Its Application to Earth Observation Satellite, doi:10. 3390/rs70303320

Chapter 6

Error Analysis of Orbit Determination for The Geostationary Satellite with Single Station Antenna Tracking Data

Chingiz Hajiyev, Melih Ata

Aeronautics and Astronautics Faculty, Istanbul Technical University, Istanbul, Turkey.

ABSTRACT

In the study, position and velocity values of a geostationary satellite are found. When performing this, a MATLAB algorithm is used for Runge-Kutta Fehlberg orbit integration method to solve spacecraft's position and velocity. Integrated method is the solution for the systems which mainly work with a single station. Method provides calculation of azimuth, elevation and range data by using the position simulation results found by RKF. Errors of orbit determination are analysed. Variances of orbit parameters are chosen as the accuracy criteria. Analysis results are the indicator of the method's accuracy

INTRODUCTION

Orbit determination accuracy improvement for geostationary satellite with single station antenna tracking data is investigated in [1]. In this study, an operational orbit determination (OD) and system for the geostationary satellite (Communication, Ocean and Meteorological Satellite (COMS)) mission requires accurate satellite positioning data to accomplish image navigation registration on the ground. Ranging and tracking data, which is provided by a single ground station, is used to determine the orbit of geostationary satellite in normal operation. However, the orbital longitude of geostationary satellite is so close to that of satellite tracking sites that geometric singularity affects observability [2]. Applying an estimated azimuth bias using the ranging and tracking data provided by two stations is a method to solve the azimuth bias of a single station in singularity. Velocity increments of a wheel offloading maneuver, which is performed twice a day, are fixed by planned values without considering maneuver efficiency during OD. Using only single-station data with the correction of the azimuth bias, OD is succeed to achieve three-sigma position accuracy on the order of 1.5 km root-sum square.

In [3] comparison of Extended Kalman Filter (EKF) and Unscented Kalman Filter (UKF) for spacecraft localization via angle measurements [3] is performed. In the study, performances of two nonlinear estimators are compared for the localization of a spacecraft. It is assumed that range measurements are not available, and the localization problem is tackled on the basis of angle-only measurements. A dynamic model of the spacecraft accounting for several perturbing effects, such as Earth and Moon gravitational field asymmetry and errors associated with the Moon ephemerides, is employed. The measurement process is based on elevation and azimuth of Moon and Earth with respect to the spacecraft reference system. Position and velocity of the spacecraft are estimated using both the extended Kalman filter (EKF) and the unscented Kalman filter (UKF). The behaviour of the filters is compared on two sample missions: Earth-to-Moon transfer and geostationary orbit raising.

Localization of spacecraft is usually very accurate when GPS measurements are available [4]. The problem becomes more challenging when GPS signals are not available, like in high-Earth orbits or in long range missions such as Earth-to-Moon transfers. In these cases spacecraft navigation is often handled by ground-based tracking stations, thus making it unfeasible for low-cost spacecraft missions. In order to make spacecraft fully autonomous, it is necessary to devise self-localization and navigation algorithms relying on measurements provided by onboard sensors. In the study, the problem of spacecraft self-localization is addressed using angular measurements. First a dynamic model of the spacecraft is formulated, which takes into account several perturbing effects such as Earth and Moon gravitational field asymmetry and errors associated with the Moon ephemerides. It is assumed that the navigation system is able to estimate the spacecraft attitude (by using a star tracker sensor), and the spacecraft is equipped with line-of-sight sensors providing measurements of elevation and azimuth of Moon and Earth with respect to the spacecraft reference system. Range measurements, which are often difficult to obtain or are not sufficiently reliable, are not required. Then, position and velocity of the spacecraft are estimated by employing both the classical extended Kalman filter and the recently developed unscented Kalman filter [5]. Comparisons between EKF and UKF have been proposed in several contexts, ranging from target tracking [6], to positioning systems and virtual reality. The filters have been tested on simulated data concerning two different missions. The resulting localization errors [7] and the associated confidence intervals show that the proposed algorithms provide reliable estimates, whose accuracy is sufficient for autonomous navigation in the considered class of missions. In general, for the orbit determination purpose the Kalman filtering technique is used. Accuracy of the Kalman filter depends on the accuracy of the measurement devices significantly. Therefore, it is important to investigate the improvement of the accuracy of measurement instruments.

In the study, a geostationary satellite's orbit is determined and error analysis of the orbit determination is performed. Runge-Kutta Fehlberg Orbit Integration Method is applied to equation of motion of a satellite. Position and velocity equations are provided via RKF application. Position and velocity are calculated in 5000 points with 20 seconds time interval between every step. Application of the integrated method to the results is the most important subject of this study. It is necessary to be careful about the transformation of ECI coordinates to topocentric coordinates. This type of transformation is applied because of the Earth's rotation. When, spacecraft's position changes at every point, ground station's coordinates also change due to Earth's spin. Therefore, ground station's coordinates are calculated in 5000 points by using sidereal time and latitude. First, satellite's position is calculated in ECI coordinates. Then, it is transformed to the topocentric coordinates and range, azimuth, elevation angles are calculated. Error and variance can be analyzed after the explained procedure. Variances of the orbit parameters are chosen as

the accuracy criteria. Their changes and effects of these changes are investigated.
Study's operation concept can be defined as (**Figure 1**).

ORBIT INTEGRATION BY RUNGE-KUTTA FEHLBERG METHOD

The equation of motion of a satellite is a second order vector differential equation (see Appendix A).

Method can be applied for orbit integration. The solution of first order differential equation system can be obtained according to Babolian (1994) [8]. The following algorithm is very similar to the described algorithm of RKF. However, this algorithm includes many coefficients to be computed. 36 coefficients are calculated in every step of the 5000 iterations.

First, the initial step size h is selected, and then the RKF coefficients (Appendix B., Equation B.4) are calculated via initial conditions [9].
Position and velocity of the satellite are calculated by the following formulas (1), after obtaining the coefficients [9].

$$
\begin{aligned}
x_{n+1} &= x_n + 25(k11)/216 + 1408(k31)/2565 \\
&\quad + 2197(k41)/4104 - (k51)/5 \\
y_{n+1} &= y_n + 25(k12)/216 + 1408(k32)/2565 \\
&\quad + 2197(k42)/4104 - (k52)/5 \\
z_{n+1} &= z_n + 25(k13)/216 + 1408(k33)/2565 \\
&\quad + 2197(k43)/4104 - (k53)/5 \\
\dot{x}_{n+1} &= \dot{x}_n + 25(k14)/216 + 1408(k34)/2565 \\
&\quad + 2197(k44)/4104 - (k54)/5 \\
\dot{y}_{n+1} &= \dot{y}_n + 25(k15)/216 + 1408(k35)/2565 \\
&\quad + 2197(k45)/4104 - (k55)/5 \\
\dot{z}_{n+1} &= \dot{z}_n + 25(k16)/216 + 1408(k36)/2565 \\
&\quad + 2197(k46)/4104 - (k56)/5
\end{aligned}
\tag{1}
$$

DETERMINATION OF THE ACCURACY OF THE SPACECRAFT'S COORDINATES

Determination of the Satellite's Coordinates by Single Station Antenna Tracking Data

Measurement of angle and measurement of range methods are used at the same time in the integrated method [10]. Integrated method is usually used in radiolocation systems and it determines D range to the satellite, azimuth angle α and elevation angle β. When this method is used, the coordinates of the satellite is determined as the intersection point of the sphere state surface (D = constant), cone state surface (β = constant) and vertical plane suitable for α = constant state surface. Satellite coordinates are determined with single antenna (ground station) by integrated method. Method does not need any difficult calculations.

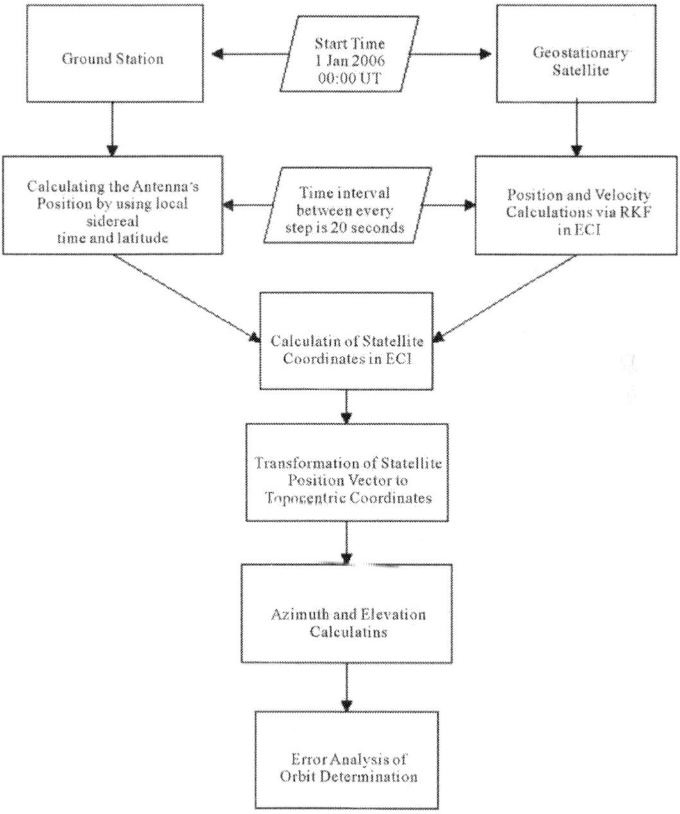

Figure 1. Operation diagram.

Advantages of the Integrated Method:

- Single station is sufficient to measure the coordinates.
- Coordinates are calculated as simple.
- Method provides the needed accuracy in whole measurement intervals.
- Information processing is easy in system.

Following formulas are used to calculate satellite coordinates;

$$x = \mathrm{D}\cos\beta\cos\alpha$$
$$y = \mathrm{D}\cos\beta\sin\alpha$$
$$z = \mathrm{D}\sin\beta \tag{2}$$

where D is the range of the satellite, α is the azimuth angle and β is the elevation angle, x,y,z are the coordinates of the range between the station and satellite The method is applied to position and velocity calculated by RKF method. The procedure is followed as shown below: The station must be denoted to satellite vector of the topocentric frame using a coordinate transformation [11,12].

The total range between satellite and ground station;

$$\mathrm{D} = C_{Ecef}^{Topocentric}\sqrt{\left(x - x_{antenna}\right)^2 + \left(y - y_{antenna}\right)^2 + \left(z - z_{antenna}\right)^2} \tag{3}$$

where;

$C_{Ecef}^{Topocentric}$ is the transformation matrix from ECEF coordinates to topocentric coordinates.

$x_{antenna}$ is the x coordinate of the antenna's position.

$y_{antenna}$ is the y coordinate of the antenna's position.

$z_{antenna}$ is the z coordinate of the antenna's position.

Calculation of azimuth and elevation angles by total range and range components;

$$\beta = \arcsin\left(\frac{z}{D}\right)$$

$$\alpha = \arctan\left(\frac{y}{x}\right) \quad (4)$$

Figures which present range, azimuth and elevation simulations are distributed in Figures 4 and 5.

Variance Analysis of the Satellite's Coordinates

Characteristics of the position data error depend on the type of ground station, location of ground station, and measurement accuracy of navigation parameters [10]. It is possible to establish ground stations optimally and choose the most accurate region to determine range via necessary calculations. However, calculations are sometimes difficult and some great preparations are necessary. In this case, approximate accuracy values of the coordinates are used to determine the accuracy of range data. Because, x, y, z coordinates of the spacecraft's range are non-linear function of three parameters measured by angle-range measurement method, the variances of coordinates' calculation errors are determined as (Equation (5));

$$\sigma_x^2 = \sum_i \left(\frac{\delta x}{\delta a_i}\right)^2 \sigma_{a_i}^2;$$

$$\sigma_y^2 = \sum_i \left(\frac{\delta y}{\delta a_i}\right)^2 \sigma_{a_i}^2;$$

$$\sigma_z^2 = \sum_i \left(\frac{\delta z}{\delta a_i}\right)^2 \sigma_{a_i}^2; \quad (5)$$

where σ_x^2, σ_y^2 and σ_z^2 are variances of the calculated coordinates' errors. $\sigma_{a_i}^2$ is the variances of the navigation parameters' measurement errors.

Let's derive the equations which evaluate the accuracy of the coordinates determined via integrated method [10]. The coordinates of the spacecraft are calculated by formulas; $x = D\cos\beta\cos\alpha$, $y = D\cos\beta\sin\alpha$, $z = D\sin\beta$ in this method.

Some equations can be written by assumption of that navigation parameters' measurement errors are independent;

$$\sigma_x^2 = \left(\frac{\delta x}{\delta \mathrm{D}}\right)^2 \sigma_D^2 + \left(\frac{\delta x}{\delta \beta}\right)^2 \sigma_\beta^2 + \left(\frac{\delta x}{\delta \alpha}\right)^2 \sigma_\alpha^2;$$

$$\sigma_y^2 = \left(\frac{\delta y}{\delta \mathrm{D}}\right)^2 \sigma_D^2 + \left(\frac{\delta y}{\delta \beta}\right)^2 \sigma_\beta^2 + \left(\frac{\delta y}{\delta \alpha}\right)^2 \sigma_\alpha^2;$$

$$\sigma_z^2 = \left(\frac{\delta z}{\delta \mathrm{D}})\right)^2 \sigma_D^2 + \left(\frac{\delta z}{\delta \beta})\right)^2 \sigma_\beta^2; \qquad (6)$$

σ_D^2, σ_β^2, σ_α^2 are variances of measurement errors.

$$\frac{\delta x}{\delta \mathrm{D}} = \cos\beta\cos\alpha; \frac{\delta x}{\delta\beta} = -\mathrm{D}\cos\alpha\sin\beta;$$

$$\frac{\delta x}{\delta\alpha} = -\mathrm{D}\cos\beta\sin\alpha; \frac{\delta y}{\delta \mathrm{D}} = \cos\beta\sin\alpha;$$

$$\frac{\delta y}{\delta\beta} = -\mathrm{D}\sin\alpha\sin\beta; \frac{\delta y}{\delta\alpha} = \mathrm{D}\cos\beta\cos\alpha;$$

$$\frac{\delta z}{\delta \mathrm{D}} = \sin\beta; \frac{\delta z}{\delta\beta} = \mathrm{D}\cos\beta; \frac{\delta z}{\delta\alpha} = 0; \qquad (7)$$

Spherical standard deviation is used as the value of the accuracy of spacecraft's position. It can be written as below;

$$\sigma_k = \sqrt{\sigma_x^2 + \sigma_y^2 + \sigma_z^2}$$

$$= \sqrt{\sigma_D^2 + \mathrm{D}^2\sigma_\beta^2 + \mathrm{D}^2\cos^2\beta\sigma_\alpha^2} \qquad (8)$$

SIMULATIONS

Satellite's initial position and velocity data are distributed (**Table 1**). Firstly, satellite's velocity and position values are calculated in 5000 points. It is done by Runge-Kutta Fehlberg Orbit Integration method. Then, range, azimuth and elevation angle data are obtained via integrated method. Their new values are calculated by addition of some noises and standard deviations. Errors between initial range, azimuth and elevation values and new values are determined and their standard deviation is analyzed. Finally, variance analysis and spherical standard deviations of the results are found and the results are compared. Initial conditions are shown as (**Table 1**).

Satellite is at 128.2 deg east longitude and the starting time for observation is 1 Jan 2006, 00:00 UT.

Single station antenna which is used for measurements has position data distributed as (**Table 2**).

1) Simulations for Position and Velocity

The position and velocity values are calculated in 5000 points (**Table 3**) with initial conditions in**Table 1** and the step size h = 20 by calculating RKF coefficients (Equation B.4). Their simulations are also distributed in Figures 2 and 3.

WhereXrk is the x position coordinate of the satellite.

Yrk is the y position coordinate of the satellite.

Zrk is the z position coordinate of the satellite.

Vxrk is the x velocity coordinate of the satellite.

Vyrk is the y velocity coordinate of the satellite.

Vzrk is the z velocity coordinate of the satellite.

Velocity simulations for x, y and z coordinates are shown in **Figure 2**.

Position simulations for x, y and z coordinates are shown in **Figure 3**.

2) Simulations for Azimuth, Elevation and Range

The simulation of the range between spacecraft and antenna is shown in **Figure 4**.

As it can be seen, range generally decreases up to 2500th step. Then, it shows an increase. For example:

Table 1. Initial position and velocity value.

Position	X(m)	Y(m)	Z(m)
Initial Conditions	−27,829,408.38	−31,684,615.8	3477.99
Velocity	**Vx (m/s)**	**Vy (m/s)**	**Vz (m/s)**
Initial Conditions	2309.78	−2028.7	−1.92

Table 2. Location of ground station.

Position	Longitude (deg)	Latitude (deg)	e
Ground Station Antenna	127.36 east	36.4 north	0.00014142

Table 3. Position and velocity values

Values	500	1000	1500	2000	2500	3000	3500	4000	4500	5000
Xrk (m)	2.88E + 05	2.83E + 07	4.19E + 07	3.42E + 07	9.11E + 06	−2.06E + 07	−3.99E + 07	−3.89E + 07	−1.81E + 07	1.19E + 07
Yrk (m)	−4.22E + 07	−3.13E + 07	−4.44E + 06	2.46E + 07	4.12E + 07	3.68E + 07	1.37E + 07	−1.64E + 07	−3.81E + 07	−4.05E + 07
Zrk (m)	−1.49E + 04	−2.58E + 04	−2.35E + 04	−9.30E + 03	9.65E + 03	2.37E + 04	2.57E + 04	1.46E + 04	−3.88E + 03	−2.04E + 04
Vxrk (m/s)	3.07E + 03	2.28E + 03	323.8212	−1.80E + 03	−3.00E + 03	−2.68E + 03	−996.689	1.19E + 03	2.78E + 03	2.95E + 03
Vyrk(m/s)	21.3198	2.06E + 03	3.06E + 03	2.50E + 03	664.3382	−1.51E + 03	−2.91E + 03	−2.83E + 03	−1.32E + 03	868.3936
Vzrk (m/s)	−1.6024	−0.4704	0.9012	1.8146	1.8049	0.8771	−0.4965	−1.6174	−1.9159	−1.2403

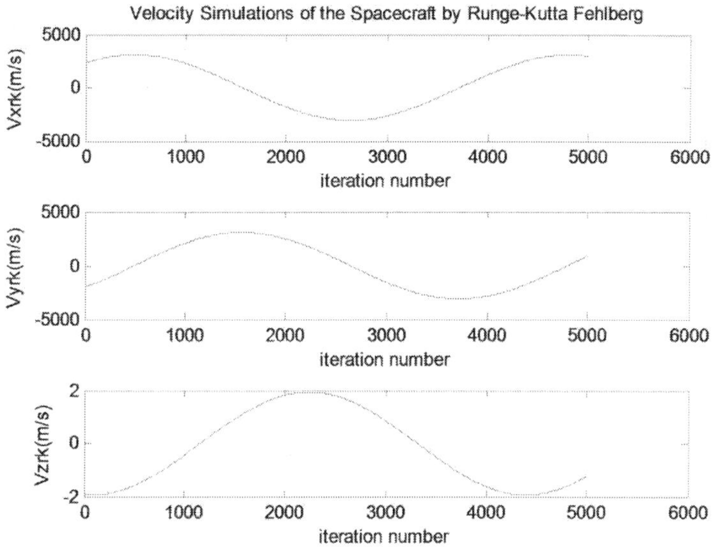

Figure 2. Velocity simulations of the satellite.

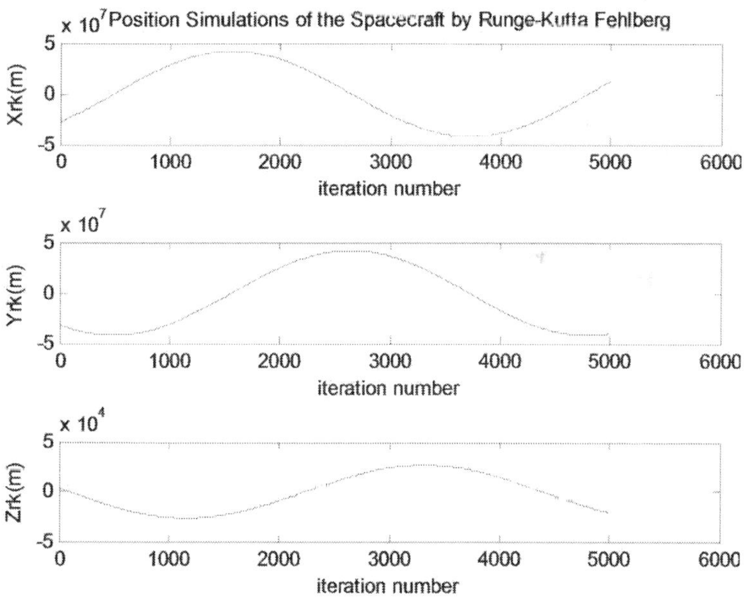

Figure 3. Position simulations of the satellite.

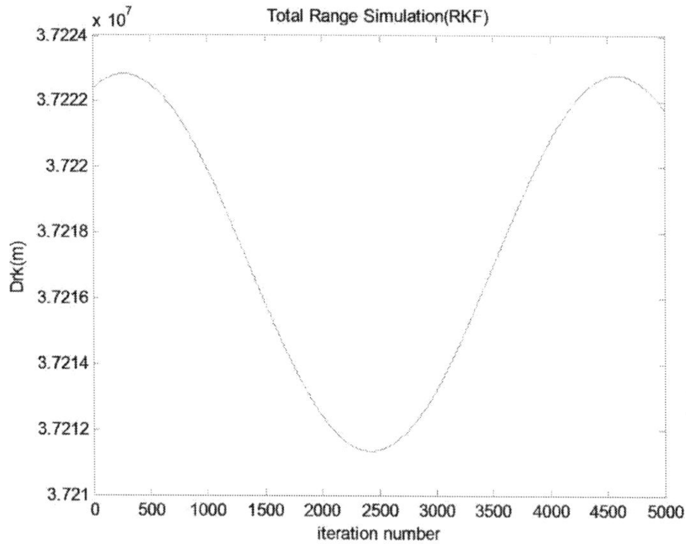

Figure 4. Total range simulation.

think the point which symbolizes the time which is a day later from starting point. This point is between the 4000th - 4500th step interval. If this point is examined, range is nearly the same with the starting point. It can be said that, range is nearly constant for a day period for geostationary satellites.

Simulations for azimuth and elevation angles are shown in **Figure 5**.
There is a little differentiation from the initial value in both of the angles. These changes are insignificant changes.

3) Simulations for Variance Analysis

The variances, variances components' analysis and spherical standard deviation results according to integrated method results applied to RKF are shown in **Table 4**. Their graphics are also shown in Figures 6-10.
The value of the mean spherical standard deviation is 7.83E + 03.
WhereVARxrk is the variance of x position coordinate.

VARyrk is the variance of y position coordinate.

VARzrk is the variance of z position coordinate.

VARxDrk is the variance of x coordinate error due to range uncertainties.

VARxbetark is the variance of x coordinate error due to elevation angle uncertainties.

VARxalfark is the variance of x coordinate error due to azimuth angle uncertainties.

VARyDrk is the variance of y coordinate error due to range uncertainties.

VARybetark is the variance of y coordinate error due to elevation angle uncertainties.

VARyalfark is the variance of y coordinate error due to azimuth angle uncertainties.

VARzDrk is the variance of z coordinate error due to range uncertainties.

VARzbetark is the variance of z coordinate error due to elevation angle uncertainties.

SP.SDEVrk is the spherical standard deviation of the position.

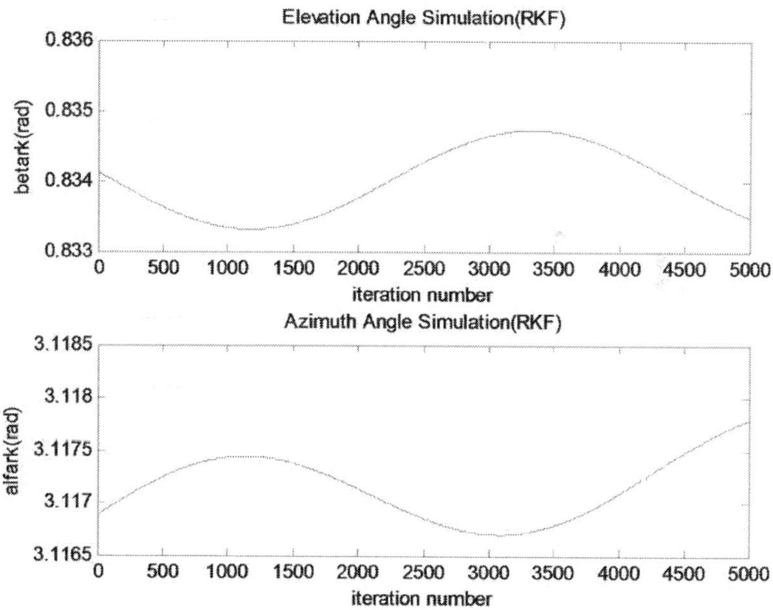

Figure 5. Angles simulation.

Table 4. Variance analysis

Values	500	1000	1500	2000	2500	3000	3500	4000	4500	5000
VARxrk (m^2)	2.31E + 07	2.31E + 07	2.31E + 07	2.31E + 07	2.31E + 07	2.32E + 07	2.32E + 07	2.32E + 07	2.31E + 07	2.31E + 07
VARyrk (m^2)	1.91E + 07	1.91E + 07	1.91E + 07	1.91E + 07	1.90E + 07	1.90E + 07	1.90E + 07	1.90E + 07	1.91E + 07	1.91E + 07
VARzrk (m^2)	1.91E + 07	1.91E + 07	1.91E + 07	1.91E + 07	1.90E + 07	1.90E + 07	1.90E + 07	1.90E + 07	1.91E + 07	1.91E + 07
VARxDrk (m^2)	45.1566	45.1866	45.1816	45.1441	45.0933	45.0549	45.0486	45.0776	45.1271	45.172
VARxbetark (m^2rad^2)	2.31E + 07	2.31E + 07	2.31E + 07	2.31E + 07	2.31E + 07	2.32E + 07	2.32E + 07	2.32E + 07	2.31E + 07	2.31E + 07
VARxalfark (m^2rad^2)	1.13E + 04	1.11E + 04	1.12E + 04	1.14E + 04	1.16E + 04	1.18E + 04	1.17E + 04	1.14E + 04	1.11E + 04	1.08E + 04
VARyDrk (m^2)	0.0268	0.0264	0.0265	0.027	0.0276	0.0279	0.0277	0.0271	0.0263	0.0256
VARybetark (m^2rad^2)	1.37E + 04	1.35E + 04	1.36E + 04	1.38E + 04	1.42E + 04	1.43E + 04	1.43E + 04	1.39E + 04	1.35E + 04	1.31E + 04
VARyalfark (m^2rad^2)	1.91E + 07	1.91E + 07	1.91E + 07	1.90E + 07	1.90E + 07	1.90E + 07	1.90E + 07	1.90E + 07	1.90E + 07	1.91E + 07
VARzDrk (m^2)	54.8166	54.787	54.7919	54.8288	54.8791	54.9172	54.9236	54.8953	54.8467	54.8024
VARzbetark (m^2rad^2)	1.91E + 07	1.91E + 07	1.91E + 07	1.91E + 07	1.90E + 07	1.90E + 07	1.90E + 07	1.90E + 07	1.91E + 07	1.91E + 07
SP.SDEVrk (m)	7.83E + 03	7.83E + 03	7.83E + 03	7.83E + 03	7.82E + 03	7.82E + 03	7.82E + 03	7.83E + 03	7.83E + 03	7.83E + 03

Variances for x, y and z positions are distributed in **Figure 6**.

Variances simulations depend on the change of the spacecraft's position coordinates. Variance changes in only small intervals on orbit operations. As it can be seen, x variance has the biggest value. The biggest amount of change (between peak point and minimum point) is around 10 km for all of the x, y and z variances. Change

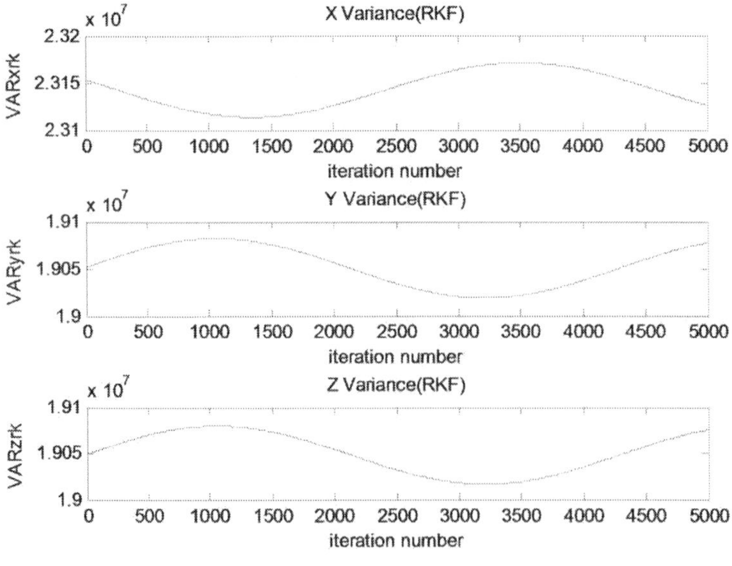

Figure 6. Variances of coordinates x, y, z.

of variances depend on the satellite's on-orbit action. These changes can take significant values for some parameters such as range, azimuth and elevation. This is an important output since the statistical characteristics of orbit determination errors are generally assumed to be constant. Components of x coordinate variance are shown in **Figure 7**.

The components of x, y and z variances are also examined to find how range, azimuth and elevation elements affect the results. It is obvious that elevation angle affects mostly the X variances. It has higher values than the elevation and range values. Range component has the weakest effect on variation.

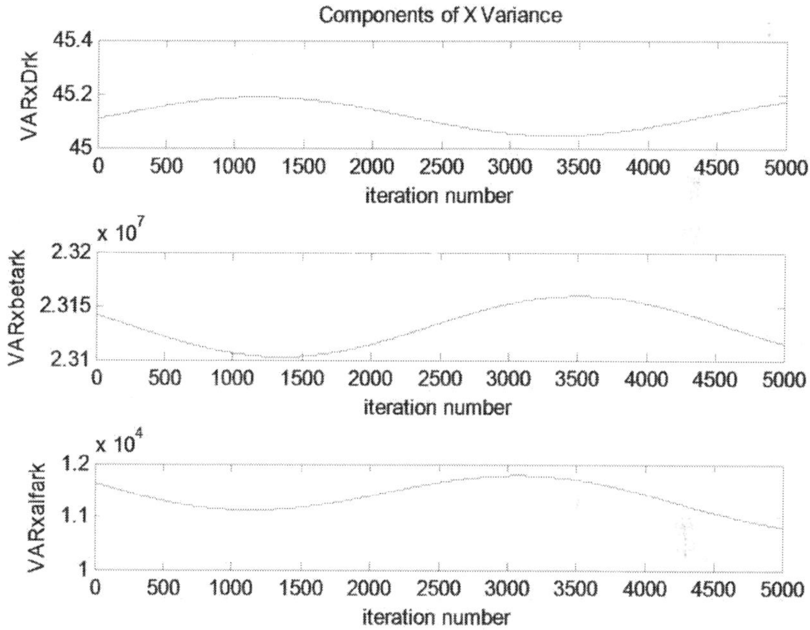

Figure 7. Components of Variance of x coordinate.

Components of y coordinate variance are shown in **Figure 8**.

Components of Y variance are investigated to understand their effects. Azimuth angle is more dominant in the Y variance. It has higher values than the elevation and range values. Range component has the weakest effect on variation. The components have insignificant variations. Components of z coordinate variance are shown in **Figure 9**.

Components of Z variance are investigated to understand their effects. Elevation angle is effective in Z variance. It has higher range value. Z variance components also have some insignificant changes.

Spherical standard deviation of the position is shown in **Figure 10**.
Spherical standard deviation changes according to the spacecraft's position. It has some insignificant changes. The amount of the biggest change is around 5 meters. It makes a peak around 7828 m and its minimum value is around 7823 meters.

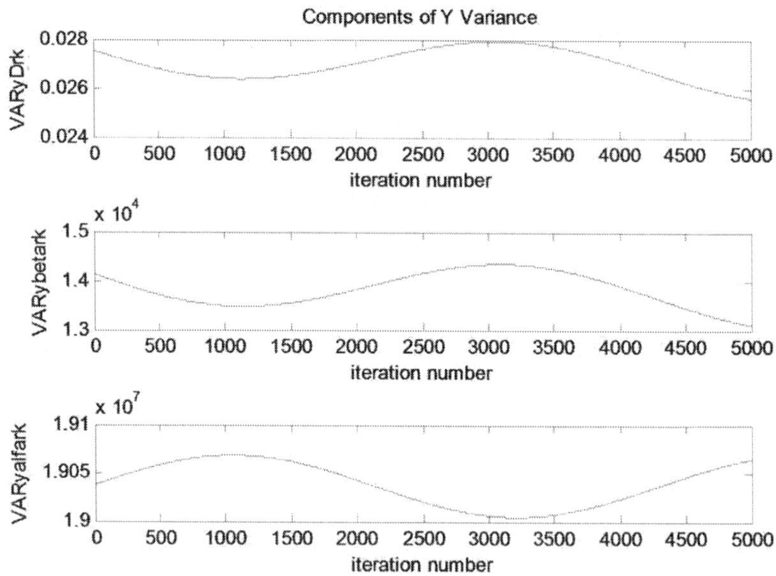

Figure 8. Components of variance of Y coordinate.

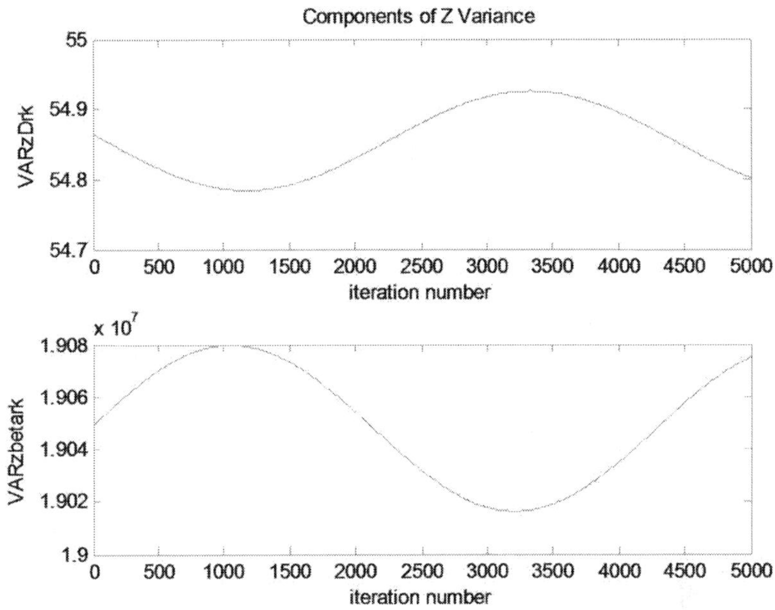

Figure 9. Components of variance of Z coordinate.

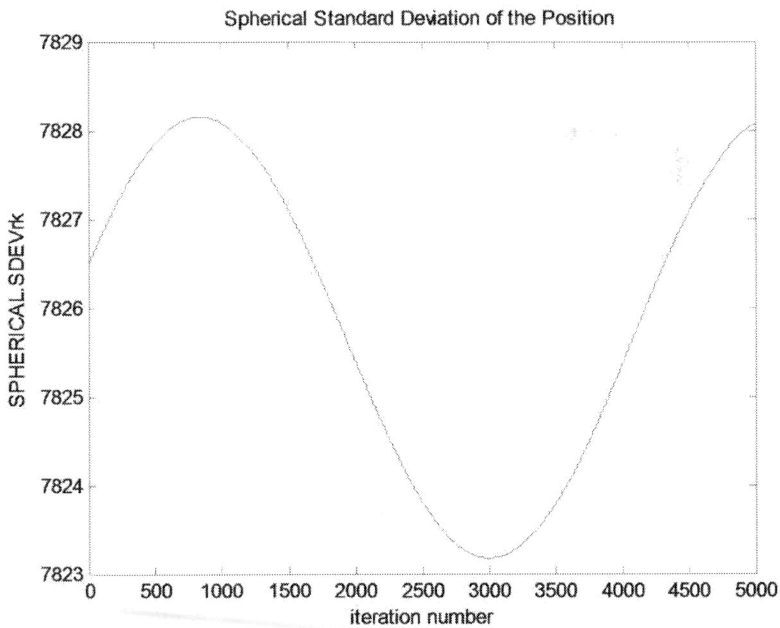

Figure 10. Spherical standard deviation of the satellite position.

CONCLUSIONS

In this study, the errors of orbit determination of geostationary satellite with single station antenna tracking data are analyzed. Variances of orbit parameters are chosen as the accuracy criteria. The orbit determination is evaluated via RKF.

Range is nearly constant for a day period of geostationary satellites. There are some small changes when compared with the initial value in azimuth and elevation angles. They changes insignificantly. It can be said that the RKF method gives accurate results by means of logical measurement results for geostationary satellite.

The components of x, y and z variances are also examined to find how range, azimuth and elevation elements affect the results. It is obvious that elevation angle affects mostly the X variances. It has higher values than the elevation and range values. Except elevation, azimuth angle is more dominant in the Y variance. Elevation angle is effective in Z variance. Generally, variances remain nearly constant in results. Range component effect is very small. They have the weakest effects on variations. Spherical standard deviation also shows some insignificant changes. According to very small changes, it can be thought as constant.

REFERENCES

1. Y. Hwang, et al., "Orbit Determination Accuracy Improvement for Geostationary Satellite with Single Station Antenna Tracking Data," ETRI Journal, Vol. 30, No. 6, 2008, pp. 774-782.
2. Y. Hwang, et al., "Communication, Ocean, and Meteorological Satellite Orbit Determination Analysis Considering Maneuver Scheme," AIAA/AAS Astrodynamics Specialist Conference and Exhibit, Keystone, Colorado, August 2006, AAS Paper 2006-6668.
3. A. Giannatrapani, et al., "Comparison of EKF and UKF for Spacecraft Localization via Angle Measurements," IEEE Transactions on Aerospace and Electronic Systems, Vol. 1, No. 47, 2011, pp. 75-84.
4. T. Upadhyay, S. Cotterill and A. W. Deaton, "Autonomous GPS/INS Navigation Experiment for Space Transfer Vehicle," IEEE Transactions on Aerospace and Electronic Systems, Vol. 3, No. 29, 1993, pp. 782-785.

5. S. J. Julier and J. K. Uhlmann, "A New Extension of the Kalman Filter to Nonlinear Systems," The 11th International Symposium on Aerospace/Defence Sensing, Simulation and Controls, Orlando, April 1997, pp. 182-193.
6. R. Zhan and J. Wan, "Iterated Unscented Kalman Filter for Passive Target Tracking," IEEE Transactions on Aerospace and Electronic Systems, Vol. 3, No. 38, 2007, pp. 1155-1163.
7. N. Ceccarelli, et al., "Spacecraft Localization via Angle Measurements for Autonomous Navigation in Deep Space," Proceedings of the 17th IFAC Symposium on Automatic Control in Aerospace, Toulouse, June 2007.
8. E. Babolian, "Numerical Methods," Elm va Sanat University Publication, Department of Mathematics, Theran, 1994.
9. M. Es-hagh, "Step Variable Numerical Orbit Integration of a Low Earth Orbiting Satellite," Journal of the Earth & Space Physics, Vol. 1, No. 31, 2005, pp. 1-12.
10. C. Haciyev, "Radyo Navigasyon," Istanbul Technical University Press, Istanbul, 1999.
11. A. Vallado, "Fundamentals of Astrodynamics and Applications," 2nd Edition, Microcosm Press Jointly with Kluwer Academic Publisher, Torrance, 2001.
12. H. D. Curtis, "Orbital Mechanics for Engineering Students," Elsevier Butterworth-Heinemann, Burlington, 2005

CITATION

C. Hajiyev and M. Ata, "Error Analysis of Orbit Determination for the Geostationary Satellite with Single Station Antenna Tracking Data," Positioning, Vol. 2 No. 4, 2011, pp. 135-144. doi: 10.4236/pos.2011.24013.

Chapter 7

Singular Formalism And Admissible Control Of Spacecraft With Rotating Flexible Solar Array

Lu Dongning [a,b], *Liu Yiwu* [b]

[a] Science and Technology on Intelligent Space Control Laboratory, Beijing Institute of Control Engineering, Beijing 100190, China
[b] Beijing Institute of Control Engineering, Beijing 100190, China

ABSTRACT

This paper is concerned with the attitude control of a three-axis-stabilized spacecraft which consists of a central rigid body and a flexible sun-tracking solar array driven by a solar array drive assembly. Based on the linearization of the dynamics of the spacecraft and the modal identities about the flexible and rigid coupling matrices, the spacecraft attitude dynamics is reduced to a formally singular system with periodically varying parameters, which is quite different from a spacecraft with fixed appendages. In the framework of the singular control theory, the regularity and impulse-freeness of the singular system is analyzed and then admissible attitude controllers are designed by Lyapunov's method. To improve the robustness against system uncertainties, an H_∞optimal control is designed by optimizing the H_∞ norm of the system transfer function matrix. Comparative numerical experiments are performed to verify the theoretical results.

INTRODUCTION

In order to maximize the efficiency of solar electric power generation, spacecraft nowadays are usually equipped with solar array drive assemblies (SADAs) to keep their solar arrays continuously facing the Sun. However, system dynamics and control of the spacecraft with movable flexible appendages become more complex than those with fixed appendages. Without loss of generality, the attitude control of a slightly complex spacecraft which consists of a central rigid body and a flexible solar array rotating slowly with an orbital angular velocity along the pitch axis, will be considered in this paper. Numerous spacecraft can be treated as this sort of system, such as the satellite examples in the Engineering Test Satellite (ETS) series of Japan, the Geostationary Operational Environmental Satellite (GOES) series of US, the Indian Remote-Sensing (IRS) series of India, and the Satellite Pour l'Observation de la Terre (SPOT) series of France. Due to the rotation of the solar array, these spacecraft are typical linear parameter varying models in which the system modal frequencies, the inertia matrix of the spacecraft, and the coupling matrices between the spacecraft attitude motions and solar array vibrations and rotation vary continuously. Generally speaking, none of these problems will arise in spacecraft with fixed appendages. One of the feasible solutions to this kind of control problem is to choose a particular position of the solar array, e.g., 45°, as the worst-case configuration and design robust controllers, such as the loop-shapingH_∞ controller for satellite SPOT-4. [1] This controller was synthesized using the H_∞coprime factorization approach and its robustness against parameter uncertainties was checked by mixed real/complex μ analysis. Using a cone complementarity linearization algorithm, low-order H_∞ controllers were developed such that the control scheme may be more attractive for practical use. [2] An alternative solution is to decompose this time-varying three-axis-coupled multi-input multi-output control system into three decoupled time-invariant single-input single-output subsystems, in which dynamic equations are transformed into solar array-fixed reference frames from body-fixed frames, [3], [4] and [5]and then design robust controllers such as linear-quadratic-Gaussian (LQG) and H_∞. This method makes use of the property that the reference frame of the sun-tracking solar array is actually an inertial frame. However, as was pointed out in Ref. [6], this decoupling technique is an approximation which requires that the inertia around the roll/yaw axes of the central rigid body be symmetrical. With the development of onboard computational capabilities, various sophisticated control algorithms in the framework of gain scheduling are studied. However, scheduling algorithms of control gains may be time and memory consuming. Thus Ref. [7] developed a linear interpolation law of the gains such that the gain-scheduling algorithm can be easily implemented in an onboard satellite computer. As reported in Ref. [8], the gain scheduling control scheme has been tested by ETS-VIII attitude control experiments on-orbit and compared with other control approaches such as the classical PID control. However, generally almost all of the control strategies mentioned above neglected the rotation dynamics of the solar array.

Different from the above work, this paper will reduce the spacecraft model to a singular system by taking into account the dynamics of the rotating flexible solar array. Letting$\boldsymbol{x} \in \mathbf{R}n$ ($n \geq 2$) denote a state vector and $\boldsymbol{f}$: $\mathbf{R}n \times \mathbf{R}n \times \mathbf{R}_{+} \rightarrow \mathbf{R}n$ denote a vector function, the mathematical model of the so-called singular systems is of the following form:

$$\boldsymbol{M}(t)\ddot{\boldsymbol{x}}(t) = \boldsymbol{f}(\dot{\boldsymbol{x}}, \boldsymbol{x}, t) \tag{1}$$

where the coefficient matrix $\boldsymbol{M}(t)$ of the acceleration term is denominated as the generalized mass matrix, and det $\boldsymbol{M}(t) \equiv 0$. Singular systems are also called differential-algebraic equation systems since some algebraic constraints can be found explicitly in Eq. (1) besides differential equations. Thus some of the state variables in singular systems are redundant and should in principle be removed. However, when redundant variables are of practical interest and importance (like the rotation and flexible vibrations of the appendage in the spacecraft system under consideration), they should not be simply canceled. Moreover, redundant variables could not be derived explicitly if algebraic constraints are transcendental equations. On the other hand, some physical phenomena such as impulse and non-causality can be modeled and treated properly in the framework of a singular system instead of the classical state-space system. [9] As a natural and convenient modeling technique and a generalized control method, the singular system theory has been applied to large scale systems, economic processes, electrical networks, and complex mechanical systems, etc.

This paper is organized as follows. In Section 2, the dynamics and kinematics of the spacecraft are discussed in detail. Based on the modal identities, mathematical singularity of the spacecraft model is investigated and a generalized state-space formulation is presented in

Section 3. In Section 4, an analysis in terms of the regularity and impulse-free property of the system is performed in the framework of singular control theory. On this basis, admissible attitude control laws with output feedback are constructed by Lyapunov's method. Then an H_∞ optimal control is introduced to improve the system robustness against the model and variable uncertainties. Numerical simulations are carried out based on the finite element analysis data of GOES-8 satellite in Section 5 to verify the controller designs and make a comparison with other controllers. Concluding remarks are made and some possible research topics are proposed in Section 6.

SPACECRAFT ATTITUDE DYNAMICS AND PROBLEM STATEMENT

Complex spacecraft usually have several large, flexible, and/or movable appendages connected to their core bodies. An accurate kinetic model of these spacecraft will contain quite a number of high-order nonlinear terms, and its system analysis and control design may become very difficult if not impossible. For tractability several assumptions are necessary to simplify the complex spacecraft model in practice, namely: (i) the orbital and the attitude motions of the spacecraft are decoupled from each other (i.e., orbital dynamics will not be considered in attitude dynamics); (ii) the center of mass of the spacecraft will not change in spite of the rotations and vibrations of the solar array; and (iii) variables of the attitude motion of the spacecraft, and those of the rotations and vibrations of the solar array are all first-order infinitesimal variables such that their products are high-order infinitesimal terms and can be neglected in the mathematical model.

Let $\boldsymbol{\omega} = [\omega x\ \omega y\ \omega z]^T \in \mathbf{R}^3$, $\boldsymbol{\Omega} \in \mathbf{R}$, and $\boldsymbol{\eta} = [\eta_1\ \eta_2\ \ldots\ \eta N]^T \in \mathbf{R}N$ denote respectively the angular velocities of the spacecraft, the angular velocity of the solar array, and modal coordinates with truncation number N, one of the well-known formulations of the attitude dynamics of the spacecraft system, as shown in Fig. 1, is then presented as follows[10] and [11]

$$\begin{cases} J\dot{\omega} + J_{bs}l\dot{\Omega} + H_{bs}\ddot{\eta} = T + \mathrm{HOT}_1(\omega, \Omega, \dot{\eta}, \eta) \\ l^T J_{bs}^T \dot{\omega} + l^T I_s l \dot{\Omega} + l^T H_s \ddot{\eta} = \tau + \mathrm{HOT}_2(\omega, \Omega, \dot{\eta}, \eta) \\ H_{bs}^T \dot{\omega} + H_s^T l \dot{\Omega} + \ddot{\eta} = -D\dot{\eta} - K\eta + \mathrm{HOT}_3(\omega, \Omega, \dot{\eta}, \eta) \end{cases} \tag{2}$$

where $\mathrm{HOT}_1(\omega, \boldsymbol{\Omega}, \dot{\eta}, \eta)$, $\mathrm{HOT}_2(\omega, \boldsymbol{\Omega}, \dot{\eta}, \eta)$ and $\mathrm{HOT}_3(\omega, \boldsymbol{\Omega}, \dot{\eta}, \eta)$ are the high-order infinitesimal terms that do not contain any acceleration variables $(\dot{\omega}, \dot{\boldsymbol{\Omega}}$, or $\ddot{\eta})$. $\boldsymbol{l} = [0\ 1\ 0]^T$ is the rotational direction of the solar array, which means the solar array is rotating along the pitch axis of the spacecraft. The $N \times N$ diagonal matrices $\boldsymbol{D}$ and $\boldsymbol{K}$ are the orthonormal modal damping and stiffness of the flexible appendage, and both are strictly positive definite. Note that the inertia matrix of the whole spacecraft with respect to (w.r.t.) the central body-fixed reference frame F_b is

$$\boldsymbol{J} = \boldsymbol{J}_b + \boldsymbol{J}_s \tag{3}$$

where $\boldsymbol{J}_b$ is the inertia matrix of the central body, and $\boldsymbol{J}_s$ is that of the solar array, which is expressed explicitly by [10]

$$J_s = C_{bs} I_s C_{bs}^T + m_s \tilde{b}\tilde{b}^T + \tilde{b} C_{bs} \tilde{c}_s^T C_{bs}^T + C_{bs} \tilde{c}_s C_{bs}^T \tilde{b}^T \tag{4}$$

where m_s is the total mass of the solar array and $\boldsymbol{I}_s$ is the inertia matrix of the solar array w.r.t. the solar array-fixed reference frame F_s. $\tilde{\boldsymbol{b}}$ and $\tilde{\boldsymbol{c}}_s$ are respectively the skew-symmetric matrices of $\boldsymbol{b}$ (position vector of the hinge point of the solar array) and $\boldsymbol{c}_s$(the first moment of inertia of the solar array). For example,

$$\boldsymbol{b} = [b_1 \quad b_2 \quad b_3]^T \Rightarrow \tilde{\boldsymbol{b}} = \begin{bmatrix} 0 & -b_3 & b_2 \\ b_3 & 0 & -b_1 \\ -b_2 & b_1 & 0 \end{bmatrix}$$

The coordinate transformation matrix $\boldsymbol{C}_{bs}$ transforms the vectors between reference frames from F_s to F_b:

$$C_{bs} = C_{bs}(\alpha) = \begin{bmatrix} \cos\alpha & 0 & \sin\alpha \\ 0 & 1 & 0 \\ -\sin\alpha & 0 & \cos\alpha \end{bmatrix}$$

in which $\alpha \in [0,2\pi)$ is the rotation angle of the solar array around the pitch axis.$\boldsymbol{J}_{bs} \in \mathbf{R}^{3\times3}$ is denominated as the rigid coupling matrix and has the following expression [10]:

$$\boldsymbol{J}_{bs} = \boldsymbol{C}_{bs}\boldsymbol{I}_s + \tilde{\boldsymbol{b}}\boldsymbol{C}_{bs}\tilde{\boldsymbol{c}}_s^{T} \tag{5}$$

As a counterpart, $\boldsymbol{H}_s$ and $\boldsymbol{H}_{bs}$ are two flexible coupling matrices. Let $\boldsymbol{r}j$ be the position vector of the solar array elemental mass mj, then their explicit expressions are [10]

$$\boldsymbol{H}_s = \sum_{j\in SA} m_j \tilde{\boldsymbol{r}}_j \boldsymbol{\Phi}_j \tag{6}$$

And

$$\boldsymbol{H}_{bs} = \sum_{j\in SA} m_j(\tilde{\boldsymbol{b}}\boldsymbol{C}_{bs} + \boldsymbol{C}_{bs}\tilde{\boldsymbol{r}}_j)\boldsymbol{\Phi}_j = \tilde{\boldsymbol{b}}\boldsymbol{C}_{bs}\boldsymbol{P}_s + \boldsymbol{C}_{bs}\boldsymbol{H}_s \tag{7}$$

where $\boldsymbol{P}_s=\sum_{j\in SA} m_j\boldsymbol{\Phi}_j$ and $\boldsymbol{\Phi}j$ is the jth mode shape of the flexible appendage. Clearly, $\boldsymbol{H}_s$ and $\boldsymbol{P}_s$ are constant matrices given by finite element analysis, whereas $\boldsymbol{H}_{bs}$and $\boldsymbol{J}_{bs}$ are periodically varying matrices w.r.t. α.

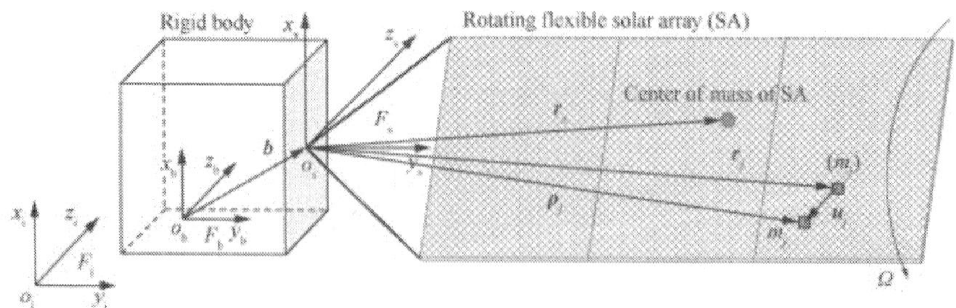

Figure. 1. Schematic of the spacecraft with a rotating flexible solar array.

Finally, the control torques are respectively $\boldsymbol{T} = [Tx\ Ty\ Tz]^{\mathrm{T}} \in \mathbf{R}^3$ generated by the reaction wheels mounted on the central bus, and $\tau \in \mathbf{R}$ generated by a driving motor inside the SADA.

The basic control purpose for this three axis stabilized spacecraft is to continuously keep the central body facing the Earth, and the solar array facing the Sun, while flexible vibrations of the appendage are damped out, i.e.,

$$\lim_{t \to \infty} \left[\boldsymbol{\omega}^{\mathrm{T}}\ \dot{\boldsymbol{\eta}}^{\mathrm{T}}\ \boldsymbol{\eta}^{\mathrm{T}}\ \boldsymbol{\Omega}\right]^{\mathrm{T}} = \left[\boldsymbol{\omega}_{\mathrm{d}}^{\mathrm{T}}\ \mathbf{0}^{\mathrm{T}}\ \mathbf{0}^{\mathrm{T}}\ \boldsymbol{\Omega}_{\mathrm{d}}\right]^{\mathrm{T}} \tag{8}$$

where the desired velocity of the spacecraft $\boldsymbol{\omega}_{\mathrm{d}}$ and that of the solar array $\boldsymbol{\Omega}_{\mathrm{d}}$constantly depend upon the orbital angular velocity $\boldsymbol{\Omega}_{\mathrm{o}}$:

$$\boldsymbol{\omega}_{\mathrm{d}} = \left[0\ \ -\boldsymbol{\Omega}_{\mathrm{o}}\ 0\right]^{\mathrm{T}}, \boldsymbol{\Omega}_{\mathrm{d}} = \boldsymbol{\Omega}_{\mathrm{o}}$$

In order to change this tracking problem to a regulating one, let $\Delta\boldsymbol{\omega}$ and $\Delta\boldsymbol{\Omega}$ be the tracking errors of the angular velocities, then they could be written as

$$\begin{cases} \Delta\boldsymbol{\omega} = \boldsymbol{\omega} - \boldsymbol{C}_{\mathrm{bd}}(\boldsymbol{q})\boldsymbol{\omega}_{\mathrm{d}} \\ \Delta\boldsymbol{\Omega} = \boldsymbol{\Omega} - \boldsymbol{\Omega}_{\mathrm{d}} \end{cases}$$

Explicitly, $\boldsymbol{C}_{\mathrm{bd}}(\boldsymbol{q})$ is of the form

$$\boldsymbol{C}_{\mathrm{bd}}(\boldsymbol{q}) = \left(q_0^2 - \boldsymbol{q}_{\mathrm{v}}^{\mathrm{T}}\boldsymbol{q}_{\mathrm{v}}\right)\mathbf{1}_{3\times3} + 2\boldsymbol{q}_{\mathrm{v}}\boldsymbol{q}_{\mathrm{v}}^{\mathrm{T}} - 2q_0\tilde{\boldsymbol{q}}_{\mathrm{v}}$$

where the identity matrix $\mathbf{1}_{3\times3} \in \mathbf{R}^{3\times3}$, and the unit quaternion parameter of the tracking error

$$\boldsymbol{q} = \left[q_0 \mid q_1\ q_2\ q_3\right]^{\mathrm{T}} = \left[q_0 \mid \boldsymbol{q}_{\mathrm{v}}^{\mathrm{T}}\right]^{\mathrm{T}}$$

satisfy the following constraint:

$$q_0^2 + \boldsymbol{q}_{\mathrm{v}}^{\mathrm{T}}\boldsymbol{q}_{\mathrm{v}} = 1$$

With the above definition of the unit quaternion, the kinematics of the attitude tracking error is given by the following equations:

$$\dot{q}_0 = -\frac{1}{2}\boldsymbol{q}_{\mathrm{v}}^{\mathrm{T}}\Delta\boldsymbol{\omega}, \quad \dot{\boldsymbol{q}}_{\mathrm{v}} = \frac{1}{2}(q_0\mathbf{1}_{3\times 3} + \tilde{\boldsymbol{q}}_{\mathrm{v}})\Delta\boldsymbol{\omega} \tag{9}$$

Meanwhile the angle error of the solar array is derived by

$$\Delta\dot{\alpha} = \boldsymbol{\Omega} - \boldsymbol{\Omega}_{\mathrm{d}} = \Delta\boldsymbol{\Omega} \tag{10}$$

Thus, the control purpose Eq. (8) has become a regulation problem, i.e., a way to develop suitable controllers $\boldsymbol{T}$ and τ via the feedback of measurable outputs $\boldsymbol{q}$, $\boldsymbol{\omega}$, α, and $\boldsymbol{\Omega}$ such that

$$\lim_{t\to\infty}\left[1 - q_0 \;\; \boldsymbol{q}_{\mathrm{v}}^{\mathrm{T}} \;\; \Delta\boldsymbol{\omega}^{\mathrm{T}} \;\; \dot{\boldsymbol{\eta}}^{\mathrm{T}} \;\; \boldsymbol{\eta}^{\mathrm{T}} \;\; \Delta\alpha \;\; \Delta\boldsymbol{\Omega}\right]^{\mathrm{T}} = \left[0 \;\; \mathbf{0}^{\mathrm{T}} \;\; \mathbf{0}^{\mathrm{T}} \;\; \mathbf{0}^{\mathrm{T}} \;\; \mathbf{0}^{\mathrm{T}} \;\; 0 \;\; 0\right]^{\mathrm{T}} \tag{11}$$

MATHEMATICAL SINGULARITY AND SINGULAR FORM

Modal Identities and Mathematical Singularity of the System

In the flexible spacecraft model, the number of flexible modal variables is usually quite large. Thus modal truncation is necessary and one of the commonly-used criteria is the so-called *inertial completeness index*, [12] which is introduced on the basis of modal identities. In the case of the fixed solar array, there exist the following well-known modal identities w.r.t. F_{s},

$$\boldsymbol{P}_{\mathrm{s}}\boldsymbol{P}_{\mathrm{s}}^{\mathrm{T}} \equiv m_{\mathrm{s}}\mathbf{1}_{3\times 3}, \quad \boldsymbol{H}_{\mathrm{s}}\boldsymbol{P}_{\mathrm{s}}^{\mathrm{T}} \equiv \tilde{\boldsymbol{c}}_{\mathrm{s}}, \quad \boldsymbol{H}_{\mathrm{s}}\boldsymbol{H}_{\mathrm{s}}^{\mathrm{T}} \equiv \boldsymbol{I}_{\mathrm{s}} \tag{12}$$

For the rotating appendage within the present context, one can obtain that

$$\boldsymbol{J} - \boldsymbol{H}_{\mathrm{bs}}\boldsymbol{H}_{\mathrm{bs}}^{\mathrm{T}} \equiv \boldsymbol{J}_{\mathrm{b}}, \quad \boldsymbol{J}_{\mathrm{bs}} - \boldsymbol{H}_{\mathrm{bs}}\boldsymbol{H}_{\mathrm{s}}^{\mathrm{T}} \equiv \mathbf{0}_{3\times 3} \tag{13}$$

Actually, by combining identities (Eq. (12)) with Eqs. (3), (4), (5) and (7), we get

$$\begin{aligned}\boldsymbol{H}_{\mathrm{bs}}\boldsymbol{H}_{\mathrm{s}}^{\mathrm{T}} &= (\tilde{\boldsymbol{b}}\boldsymbol{C}_{\mathrm{bs}}\boldsymbol{P}_{\mathrm{s}} + \boldsymbol{C}_{\mathrm{bs}}\boldsymbol{H}_{\mathrm{s}})\boldsymbol{H}_{\mathrm{s}}^{\mathrm{T}} = \tilde{\boldsymbol{b}}\boldsymbol{C}_{\mathrm{bs}}\boldsymbol{P}_{\mathrm{s}}\boldsymbol{H}_{\mathrm{s}}^{\mathrm{T}} + \boldsymbol{C}_{\mathrm{bs}}\boldsymbol{H}_{\mathrm{s}}\boldsymbol{H}_{\mathrm{s}}^{\mathrm{T}} = \tilde{\boldsymbol{b}}\boldsymbol{C}_{\mathrm{bs}}\tilde{\boldsymbol{c}}_{\mathrm{s}}^{\mathrm{T}} + \boldsymbol{C}_{\mathrm{bs}}\boldsymbol{I}_{\mathrm{s}} \\ &= \boldsymbol{J}_{\mathrm{bs}}\end{aligned}$$

And

$$\begin{aligned}\boldsymbol{H}_{\mathrm{bs}}\boldsymbol{H}_{\mathrm{bs}}^{\mathrm{T}} &= (\tilde{\boldsymbol{b}}\boldsymbol{C}_{\mathrm{bs}}\boldsymbol{P}_{\mathrm{s}} + \boldsymbol{C}_{\mathrm{bs}}\boldsymbol{H}_{\mathrm{s}})(\tilde{\boldsymbol{b}}\boldsymbol{C}_{\mathrm{bs}}\boldsymbol{P}_{\mathrm{s}} + \boldsymbol{C}_{\mathrm{bs}}\boldsymbol{H}_{\mathrm{s}})^{\mathrm{T}} \\ &= \tilde{\boldsymbol{b}}\boldsymbol{C}_{\mathrm{bs}}\boldsymbol{P}_{\mathrm{s}}\boldsymbol{P}_{\mathrm{s}}^{\mathrm{T}}\boldsymbol{C}_{\mathrm{bs}}^{\mathrm{T}}\tilde{\boldsymbol{b}}^{\mathrm{T}} + \tilde{\boldsymbol{b}}\boldsymbol{C}_{\mathrm{bs}}\boldsymbol{P}_{\mathrm{s}}\boldsymbol{H}_{\mathrm{s}}^{\mathrm{T}}\boldsymbol{C}_{\mathrm{bs}}^{\mathrm{T}} + \boldsymbol{C}_{\mathrm{bs}}\boldsymbol{H}_{\mathrm{s}}\boldsymbol{P}_{\mathrm{s}}^{\mathrm{T}}\boldsymbol{C}_{\mathrm{bs}}^{\mathrm{T}}\tilde{\boldsymbol{b}}^{\mathrm{T}} + \boldsymbol{C}_{\mathrm{bs}}\boldsymbol{H}_{\mathrm{s}}\boldsymbol{H}_{\mathrm{s}}^{\mathrm{T}}\boldsymbol{C}_{\mathrm{bs}}^{\mathrm{T}} \\ &= m_{\mathrm{s}}\tilde{\boldsymbol{b}}\tilde{\boldsymbol{b}}^{\mathrm{T}} + \tilde{\boldsymbol{b}}\boldsymbol{C}_{\mathrm{bs}}\tilde{\boldsymbol{c}}_{\mathrm{s}}^{\mathrm{T}}\boldsymbol{C}_{\mathrm{bs}}^{\mathrm{T}} + \boldsymbol{C}_{\mathrm{bs}}\tilde{\boldsymbol{c}}_{\mathrm{s}}\boldsymbol{C}_{\mathrm{bs}}^{\mathrm{T}}\tilde{\boldsymbol{b}}^{\mathrm{T}} + \boldsymbol{C}_{\mathrm{bs}}\boldsymbol{I}_{\mathrm{s}}\boldsymbol{C}_{\mathrm{bs}}^{\mathrm{T}} = \boldsymbol{J}_{\mathrm{s}} = \boldsymbol{J} - \boldsymbol{J}_{\mathrm{b}}\end{aligned}$$

Identities (13) were first reported by Ref. [10], where they were proven by Laplace transformation along with the other modal identities of different forms. It is worth noting that all the identities above hold only if the system model contains all the flexible modes (that is, $N = \infty$).

Now consider spacecraft system model (2) and note that the corresponding generalized mass matrix is

$$\boldsymbol{M}_{\alpha} = \begin{bmatrix} \boldsymbol{J} & \boldsymbol{J}_{\mathrm{bs}}\boldsymbol{l} & \boldsymbol{H}_{\mathrm{bs}} \\ \boldsymbol{l}^{\mathrm{T}}\boldsymbol{J}_{\mathrm{bs}}^{\mathrm{T}} & \boldsymbol{l}^{\mathrm{T}}\boldsymbol{I}_{\mathrm{s}}\boldsymbol{l} & \boldsymbol{l}^{\mathrm{T}}\boldsymbol{H}_{\mathrm{s}} \\ \boldsymbol{H}_{\mathrm{bs}}^{\mathrm{T}} & \boldsymbol{H}_{\mathrm{s}}^{\mathrm{T}}\boldsymbol{l} & \boldsymbol{1} \end{bmatrix} \tag{14}$$

According to modal identities (12) and (13), it can be seen that the second and the third "rows" or "columns" are linearly correlated (shown via pre-multiplying the third row by $\boldsymbol{l}^{\mathrm{T}}\boldsymbol{H}_{\mathrm{s}}$ or post-multiplying the third column by $\boldsymbol{H}_{\mathrm{s}}^{\mathrm{T}}\boldsymbol{l}$). This linear correlation will strictly hold when the number of modal variables N in the system model increases to infinity, so that

$$\lim_{N\to\infty} \det \boldsymbol{M}_{\alpha} = 0$$

This is a reasonable supposition because vibrations of continuous structures consist of infinite flexible modes, as viewed from continuum mechanics. It reminds people that if adequate flexible modes are contained in the system, then it is a stiff system, i.e., a system that is very difficult to numerically integrate by explicit fixed-step methods, such as RK4 methods. Actually, by pre-multiplying the third equation of system

(2) by $\boldsymbol{l}^{\mathrm{T}}\boldsymbol{H}_{\mathrm{s}}$, together with the second equation the following torque constraint results:

$$0 = \tau + \boldsymbol{l}^{\mathrm{T}}\boldsymbol{H}_{\mathrm{s}}\boldsymbol{D}\dot{\boldsymbol{\eta}} + \boldsymbol{l}^{\mathrm{T}}\boldsymbol{H}_{\mathrm{s}}\boldsymbol{K}\boldsymbol{\eta}$$

Which indicates that the sum of the components of all the elastic torques and damping torques along the drive axis should be equal to the drive torque. This constraint relationship suggests that system (2) is a singular system.

Generalized State-space Formulation

Let $\boldsymbol{\varphi}$ be the total tracking error of the angular velocity of the rotating solar array in the modal coordinate, i.e.,

$$\boldsymbol{\varphi} = \boldsymbol{H}_{\mathrm{bs}}^{\mathrm{T}}\Delta\boldsymbol{\omega} + \boldsymbol{H}_{\mathrm{s}}^{\mathrm{T}}\boldsymbol{l}\Delta\boldsymbol{\Omega} + \dot{\boldsymbol{\eta}} \tag{16}$$

then it follows that

$$\begin{cases} \dot{\boldsymbol{\eta}} = \boldsymbol{\varphi} - \boldsymbol{H}_{\mathrm{bs}}^{\mathrm{T}}\Delta\boldsymbol{\omega} - \boldsymbol{H}_{\mathrm{s}}^{\mathrm{T}}\boldsymbol{l}\Delta\boldsymbol{\Omega} \\ \ddot{\boldsymbol{\eta}} = \dot{\boldsymbol{\varphi}} - \boldsymbol{H}_{\mathrm{bs}}^{\mathrm{T}}\Delta\dot{\boldsymbol{\omega}} - \dot{\boldsymbol{H}}_{\mathrm{bs}}^{\mathrm{T}}\Delta\boldsymbol{\omega} - \boldsymbol{H}_{\mathrm{s}}^{\mathrm{T}}\boldsymbol{l}\Delta\dot{\boldsymbol{\Omega}} \end{cases}$$

Neglecting the high-order infinitesimal term $\left(-\dot{\boldsymbol{H}}_{\mathrm{bs}}^{\mathrm{T}}\Delta\boldsymbol{\omega}\right)$ according to the linearized modeling assumptions and substituting the above two formulae into system (2) (with another high-order infinitesimal term $\dot{\boldsymbol{C}}_{\mathrm{bd}}\boldsymbol{\omega}_{\mathrm{d}}$ neglected), then letting $\boldsymbol{x}(t) = [\Delta\boldsymbol{\omega}^{\mathrm{T}}\ \boldsymbol{\varphi}^{\mathrm{T}}\ \boldsymbol{\eta}^{\mathrm{T}}\ \Delta\boldsymbol{\Omega}]^{\mathrm{T}}$, we get a generalized state-space formulation

$$\begin{cases} \boldsymbol{M}\dot{\boldsymbol{x}}(t) = \boldsymbol{A}(\alpha)\boldsymbol{x}(t) + \boldsymbol{B}\boldsymbol{u} \\ \boldsymbol{y}(t) = \boldsymbol{C}\boldsymbol{x}(t) \end{cases} \tag{17}$$

with $\boldsymbol{u} = [\boldsymbol{T}^{\mathrm{T}}\ \tau]^{\mathrm{T}}$, $\boldsymbol{y} = [\Delta\boldsymbol{\omega}^{\mathrm{T}}\ \Delta\boldsymbol{\Omega}]^{\mathrm{T}}$, and

$$M = \begin{bmatrix} J_b^{-1}(J - H_{bs}H_{bs}^T) & 0 & 0 & J_b^{-1}(J_{bs} - H_{bs}H_s^T)l \\ 0 & 1 & 0 & 0 \\ 0 & 0 & 1 & 0 \\ l^T(J_{bs}^T - H_s H_{bs}^T) & 0 & 0 & l^T(I_s - H_s H_s^T)l \end{bmatrix} = \begin{bmatrix} 1 & 0 & 0 & 0 \\ 0 & 1 & 0 & 0 \\ 0 & 0 & 1 & 0 \\ 0 & 0 & 0 & 0 \end{bmatrix} \tag{18}$$

in which all the identity matrices **1** and the null matrices **0** are of compatible dimensions. The system matrix is divided into four blocks:

$$A(\alpha) = \left[\begin{array}{c|c} A_{11}(\alpha) & A_{12}(\alpha) \\ \hline A_{21}(\alpha) & A_{22} \end{array}\right] \tag{19}$$

where

$$A_{11}(\alpha) = \begin{bmatrix} -J_b^{-1}H_{bs}DH_{bs}^T & J_b^{-1}H_{bs}D & J_b^{-1}H_{bs}K \\ DH_{bs}^T & -D & -K \\ -H_{bs}^T & 1 & 0 \end{bmatrix} \tag{20}$$

$$A_{12}(\alpha) = \begin{bmatrix} -J_b^{-1}H_{bs}DH_s^T l \\ DH_s^T l \\ -H_s^T l \end{bmatrix} \tag{21}$$

$$A_{21}(\alpha) = \begin{bmatrix} -l^T H_s DH_{bs}^T & l^T H_s D & l^T H_s K \end{bmatrix} \tag{22}$$

and a constant scalar,

$$A_{22} = -l^T H_s DH_s^T l \tag{23}$$

Finally, the input and the output matrices are

$$B = \begin{bmatrix} J_b^{-1} & 0 \\ 0 & 0 \\ 0 & 0 \\ 0 & 1 \end{bmatrix}, \quad C = \begin{bmatrix} 1 & 0 & 0 & 0 \\ 0 & 0 & 0 & 1 \end{bmatrix}$$

It is worth pointing out that the definition of the total velocity of the solar array, i.e., Eq.(16), for deriving the first-order state-space formulation is commonly used in the research on attitude control of spacecraft with fixed flexible appendages (i.e., $\alpha \equiv \alpha_0$), where the basic assumption is that the first modal identity of Eq. (13) holds no matter how many flexible modes have been truncated (see Ref. [13] for instance). For the spacecraft under consideration, when the solar array is fixed at α_0 (thus $\boldsymbol{\Omega}_0 \equiv 0$), we can get the following non-singular system:

$$\begin{cases} \boldsymbol{J}_{\mathrm{b}}\Delta\dot{\boldsymbol{\omega}} = -\boldsymbol{H}_{\mathrm{bs}}(\alpha_0)\boldsymbol{D}\boldsymbol{H}_{\mathrm{bs}}^{\mathrm{T}}(\alpha_0)\Delta\boldsymbol{\omega} + \boldsymbol{H}_{\mathrm{bs}}(\alpha_0)\boldsymbol{D}\boldsymbol{\varphi} \\ \qquad +\boldsymbol{H}_{\mathrm{bs}}(\alpha_0)\boldsymbol{K}\boldsymbol{\eta} + \boldsymbol{T} \\ \dot{\boldsymbol{\varphi}} = \boldsymbol{D}\boldsymbol{H}_{\mathrm{bs}}^{\mathrm{T}}(\alpha_0)\Delta\boldsymbol{\omega} - \boldsymbol{D}\boldsymbol{\varphi} - \boldsymbol{K}\boldsymbol{\eta} \\ \dot{\boldsymbol{\eta}} = -\boldsymbol{H}_{\mathrm{bs}}^{\mathrm{T}}(\alpha_0)\Delta\boldsymbol{\omega} + \boldsymbol{\varphi} \end{cases}$$

This is a popular model in the research on flexible spacecraft control. However, a singular system results when the first-order state-space formalism is extended to the case of the spacecraft with rotating flexible appendages in this paper.

Now notice that an algebraic equation is obtained within the singular system (17), that is

$$0 = \tau - \boldsymbol{l}^{\mathrm{T}}\boldsymbol{H}_{\mathrm{s}}\boldsymbol{D}\boldsymbol{H}_{\mathrm{bs}}^{\mathrm{T}}\Delta\boldsymbol{\omega} + \boldsymbol{l}^{\mathrm{T}}\boldsymbol{H}_{\mathrm{s}}\boldsymbol{D}\boldsymbol{\varphi} + \boldsymbol{l}^{\mathrm{T}}\boldsymbol{H}_{\mathrm{s}}\boldsymbol{K}\boldsymbol{\eta} - \boldsymbol{l}^{\mathrm{T}}\boldsymbol{H}_{\mathrm{s}}\boldsymbol{D}\boldsymbol{H}_{\mathrm{s}}^{\mathrm{T}}\boldsymbol{l}\Delta\Omega \tag{24}$$

By substituting Eq. (16) into Eq. (24), it can be seen that the algebraic equation is equivalent to Eq. (15). This constraint defines a manifold where system (17) is running.

ADMISSIBLE CONTROL SYSTEM DESIGN

Preliminaries to Singular Systems

Singular control systems are not only stable, but they should also be *regular* and *impulse-free*. Since the spacecraft is modeled as a singular system with periodically varying parameters, some preliminary concepts about time-varying singular systems are introduced as follows.

Definition 1.
14

Consider the following singular system:

$$M\dot{x}(t) = A(t)x(t) + B(t)u$$

where rank($\boldsymbol{M}$) = $r \leq n$ = dim($\boldsymbol{x}$), and dim(·) denotes the dimensions of a vector.

(i) if there exists a scalar $\lambda \in \mathbf{C}$ such that

$$\det(\lambda M - A(t)) \neq 0, \quad \forall t \geqslant 0$$

the singular system is said to be uniformly regular; and

(ii) if the following equation holds,

$$\deg(\det(\lambda M - A(t))) = r, \quad \forall t \geqslant 0$$

where deg(·) denotes the degree of a polynomial, the singular system is said to be impulse-free.

Unfortunately, the criterion of impulse-free property is not applicable when the pair ($\boldsymbol{M}$, $\boldsymbol{A}(t)$) are abstract matrices with unspecified components and symbolic dimensions, thus Lemma 1 is needed.

Lemma 1. [15]*For the following form of the matrix pair*

$$M = \begin{bmatrix} 1 & 0 \\ 0 & 0 \end{bmatrix}, \; A(t) = \begin{bmatrix} A_{11}(t) & A_{12}(t) \\ A_{21}(t) & A_{22}(t) \end{bmatrix}$$

the singular system ($\boldsymbol{M}$,$\boldsymbol{A}(t)$) *is regular, then it is impulse-free if and only if* $\boldsymbol{A}_{22}(t)$ *is nonsingular for any* $t \geq 0$.

Definition 2. [9]A singular system is said to be admissible if it is regular, impulse-free, and stable.

Lemma 2.

[16]*For a positive semi-definite Hermitian matrix* $\mathbf{D} \in \mathbf{R}^{n \times n}$ *and any matrix* $\mathbf{X}$ *satisfying* $\mathbf{X}^{H}\mathbf{X} = \mathbf{1}_{k \times k}$, *the following inequality holds*

$$\prod_{i=1}^{k} \lambda_{n-k+i}(\boldsymbol{D}) \leqslant \det(\boldsymbol{X}^{H}\boldsymbol{D}\boldsymbol{X}) \leqslant \prod_{i=1}^{k} \lambda_i(\boldsymbol{D})$$

where $\lambda_n(\mathbf{D}) \leq \lambda_{n-1}(\mathbf{D}) \leq \cdots \leq \lambda_1(\mathbf{D})$ *are all the eigenvalues of* $\mathbf{D}$, *and* $\mathbf{X}^{H}$ *is the complex conjugate transpose of* $\mathbf{X}$.

Regularity and Impulse-Freeness

Anticipating the need in subsequent subsections, the regularity and impulse-freeness of system (17) with output-feedback are considered, and the results are summarized in the following proposition:

Theorem 1.

There exists an output feedback control

$$\boldsymbol{u} = -\boldsymbol{k}\boldsymbol{y}, \quad \boldsymbol{k} = \begin{bmatrix} k_d \mathbf{1}_{3\times 3} & \mathbf{0} \\ \mathbf{0} & \kappa_d \end{bmatrix}$$

where $k_d \geq 0$ *and* $\kappa_d \geq 0$, *such that the closed-loop singular system*

$$\boldsymbol{M}\dot{\boldsymbol{x}} = (\boldsymbol{A}(\alpha) - \boldsymbol{B}\boldsymbol{k}\boldsymbol{C})\boldsymbol{x} \tag{25}$$

is regular and impulse-free. Specifically, $k_d = \kappa_d = 0$ *implies that system (17) is regular and impulse-free in essence, and* $k_d > 0$, $\kappa_d > 0$ *implies that the closed-loop system with negative feedback will not lose the regularity and impulse-freeness.*

Proof.

Let

$$\widehat{\boldsymbol{A}} = \boldsymbol{A} - \boldsymbol{B}\boldsymbol{k}\boldsymbol{C} = \left[\begin{array}{c|c} \widehat{\boldsymbol{A}}_{11} & \widehat{\boldsymbol{A}}_{12} \\ \hline \widehat{\boldsymbol{A}}_{21} & \widehat{\boldsymbol{A}}_{22} \end{array}\right]$$

where

$$\widehat{A}_{11} = \begin{bmatrix} -k_d \mathbf{1}_{3\times3} - J_b^{-1} H_{bs} D H_{bs}^T & J_b^{-1} H_{bs} D & J_b^{-1} H_{bs} K \\ D H_{bs}^T & -D & -K \\ -H_{bs}^T & 1 & 0 \end{bmatrix}$$

$$\widehat{A}_{12} = A_{12}, \widehat{A}_{21} = A_{21}$$

and the scalar

$$\widehat{A}_{22} = -\kappa_d + A_{22} = -\kappa_d - l^T H_s D H_s^T l$$

Using the modal identities, one can obtain that

$$l^T H_s H_s^T l = l^T I_s l = I_{sy} > 0$$

where I_{sy} is the moment of inertia about the y-axis of the solar array. Then the following estimation can be drawn according to Lemma 2:

$$\lambda_{min}(D) I_{sy} \leqslant l^T H_s D H_s^T l \leqslant \lambda_{max}(D) I_{sy} \tag{26}$$

where $\lambda_{min}(\boldsymbol{D})$ is the minimum eigenvalue of modal damping matrix $\boldsymbol{D}$, and $\lambda_{max}(\boldsymbol{D})$ its maximum value, and both of them are positive scalars. Therefore,

$$l^T H_s D H_s^T l > 0$$

and

$$A_{22} < 0$$

However, (i) the modal damping of space structures is usually very light, i.e., $\lambda_{min}(\boldsymbol{D}) \leqslant \lambda_{max}(\boldsymbol{D}) \ll 1$, and (ii) the moment of inertia about y-axis is usually the least one compared with the other axes so as to reduce the drive power. Therefore A_{22} may be close to zero in the worst case according to Eq. (26). Thus, if a proper value of $\kappa_d > 0$ is chosen such that

$$\widehat{A}_{22} = -\kappa_d - l^T H_s D H_s^T l \ll 0$$

the closed-loop singular system will be strictly impulse-free, if it is regular as well according to Lemma 1—actually, by Definition 1 there exists $\lambda \in \mathbf{C}$ such that

$$\det(\lambda M - \widehat{A}) = \det \begin{bmatrix} \lambda 1 - \widehat{A}_{11} & -\widehat{A}_{12} \\ -\widehat{A}_{21} & -\widehat{A}_{22} \end{bmatrix} = \det(-\widehat{A}_{22}) \det(\lambda 1 - \widehat{A}_{11} + \widehat{A}_{12}\widehat{A}_{22}^{-1}\widehat{A}_{21}) \neq 0$$

since $\widehat{A}_{22} \neq 0$, and the matrix $-\widehat{A}_{11} + \widehat{A}_{12}\widehat{A}_{22}^{-1}\widehat{A}_{21}$ is bounded. Therefore the regularity and the impulse-freeness of the system have been verified.

Remark 1.

The above proof of the regularity and impulse-free property can be generalized to the case of three-dimensional rotations of the appendage (i.e., ***l*** is dropped in Eqs.(2) and (17)), where the key step is also the estimation of $\det\left(H_s D H_s^T\right)$. Letting λ_1, λ_2, $\lambda_3$denote respectively the three maximum eigenvalues of ***D*** and λn_{-2}, λn_{-1}, λn the three minimum ones, Eq. (26) can be rewritten as

$$0 < \lambda_n \lambda_{n-1} \lambda_{n-2} \det I_s \leqslant \det\left(H_s D H_s^T\right) \leqslant \lambda_3 \lambda_2 \lambda_1 \det I_s \tag{27}$$

which shows that $H_s D H_s^T$ is also nonsingular.

A regular and impulse-free singular system guarantees that Lyapunov's method is applicable in the controller design.

Admissible Output Feedback Control

If bounded scalar functions can be found to describe all the system dynamics and the time-derivative of the functions exists, Lyapunov's method is suitable for stability analysis and control design. With this motivation, a Lyapunov function candidate is formulated to represent the whole system (9), (10) and (17), namely

$$
\begin{aligned}
V &= k_{\mathrm{p}}\left[(1-q_0)^2+\boldsymbol{q}_{\mathrm{v}}^{\mathrm{T}}\boldsymbol{q}_{\mathrm{v}}\right]+\frac{1}{2}\kappa_{\mathrm{p}}\Delta\alpha^2+\frac{1}{2}\Delta\omega^{\mathrm{T}}\boldsymbol{J}_{\mathrm{b}}\Delta\omega+\frac{1}{2}\boldsymbol{\varphi}^{\mathrm{T}}\boldsymbol{\varphi}+\frac{1}{2}\boldsymbol{\eta}^{\mathrm{T}}\boldsymbol{K}\boldsymbol{\eta}\\
&= k_{\mathrm{p}}\left[(1-q_0)^2+\boldsymbol{q}_{\mathrm{v}}^{\mathrm{T}}\boldsymbol{q}_{\mathrm{v}}\right]+\frac{1}{2}\kappa_{\mathrm{p}}\Delta\alpha^2+\frac{1}{2}\boldsymbol{x}^{\mathrm{T}}\boldsymbol{PMx}
\end{aligned}
\tag{28}
$$

with $k_p > 0$, $\kappa_p > 0$ and $\boldsymbol{P} = \mathrm{diag}(\boldsymbol{J}_b, \mathbf{1}, \boldsymbol{K}, 1)$. Obviously, Eq. (28) is bounded if and only if all the state variables, namely, q_0, $\boldsymbol{q}_v$, $\Delta\alpha$, $\Delta\boldsymbol{\omega}$, $\Delta\boldsymbol{\Omega}$, $\boldsymbol{\varphi}$, and $\boldsymbol{\eta}$, are bounded, while the boundedness of $\Delta\boldsymbol{\Omega}$ is guaranteed by the impulse-free condition—recalling algebraic equation (24), and if the control input τ takes the form of $\tau = -\kappa_d\Delta\boldsymbol{\Omega} - \kappa_p\Delta\alpha$, the following inequality can be given

$$
\Delta\boldsymbol{\Omega} = \left(-\kappa_{\mathrm{p}}\Delta\alpha - \boldsymbol{l}^{\mathrm{T}}\boldsymbol{H}_{\mathrm{s}}\boldsymbol{D}\boldsymbol{H}_{\mathrm{bs}}^{\mathrm{T}}\Delta\omega + \boldsymbol{l}^{\mathrm{T}}\boldsymbol{H}_{\mathrm{s}}\boldsymbol{D}\varphi + \boldsymbol{l}^{\mathrm{T}}\boldsymbol{H}_{\mathrm{s}}\boldsymbol{K}\eta\right)/\left(\kappa_{\mathrm{d}} + \boldsymbol{l}^{\mathrm{T}}\boldsymbol{H}_{\mathrm{s}}\boldsymbol{D}\boldsymbol{H}_{\mathrm{s}}^{\mathrm{T}}\boldsymbol{l}\right) < \infty
$$

since $\kappa_{\mathrm{d}} + \boldsymbol{l}^{\mathrm{T}}\boldsymbol{H}_{\mathrm{s}}\boldsymbol{D}\boldsymbol{H}_{\mathrm{s}}^{\mathrm{T}}\boldsymbol{l} \neq 0$. Now consider Eqs. (9), (10), (16) and (17) and note that

$$
\boldsymbol{PM} = \boldsymbol{M}^{\mathrm{T}}\boldsymbol{P} \tag{29}
$$

then the time-derivative of the scalar function V is given by a straightforward computation as

$$
\begin{aligned}
\dot{V} &= k_{\mathrm{p}}\left[2(q_0-1)\dot{q}_0+2\boldsymbol{q}_{\mathrm{v}}^{\mathrm{T}}\dot{\boldsymbol{q}}_{\mathrm{v}}\right]+\kappa_{\mathrm{p}}\Delta\alpha\Delta\dot{\alpha}+\boldsymbol{x}^{\mathrm{T}}\boldsymbol{PM}\dot{\boldsymbol{x}}\\
&= k_{\mathrm{p}}\boldsymbol{q}_{\mathrm{v}}^{\mathrm{T}}\Delta\omega+\kappa_{\mathrm{p}}\Delta\alpha\Delta\Omega+\boldsymbol{x}^{\mathrm{T}}\boldsymbol{PAx}+\boldsymbol{x}^{\mathrm{T}}\boldsymbol{PBu}\\
&= k_{\mathrm{p}}\boldsymbol{q}_{\mathrm{v}}^{\mathrm{T}}\Delta\omega+\kappa_{\mathrm{p}}\Delta\alpha\Delta\boldsymbol{\Omega}-\dot{\boldsymbol{\eta}}^{\mathrm{T}}\boldsymbol{D}\dot{\boldsymbol{\eta}}+\Delta\omega^{\mathrm{T}}\boldsymbol{T}+\Delta\boldsymbol{\Omega}\tau
\end{aligned}
\tag{30}
$$

Clearly, if the control laws are designed as

$$
\begin{cases}
\boldsymbol{T} = -k_{\mathrm{d}}\Delta\omega - k_{\mathrm{p}}\boldsymbol{q}_{\mathrm{v}}\\
\tau = -\kappa_{\mathrm{d}}\Delta\boldsymbol{\Omega} - \kappa_{\mathrm{p}}\Delta\alpha
\end{cases}
\tag{31}
$$

Then

$$
\begin{aligned}
\dot{V} &= -\dot{\boldsymbol{\eta}}^{\mathrm{T}}\boldsymbol{D}\dot{\boldsymbol{\eta}} - k_{\mathrm{d}}\Delta\omega^{\mathrm{T}}\Delta\omega - \kappa_{\mathrm{d}}\Delta\boldsymbol{\Omega}^2\\
&\leqslant -\lambda_{\min}(\boldsymbol{D})\|\dot{\boldsymbol{\eta}}\|^2 - k_{\mathrm{d}}\|\Delta\boldsymbol{\omega}\|^2 - \kappa_{\mathrm{d}}\|\Delta\boldsymbol{\Omega}\|^2\\
&\leqslant -\sigma\|\boldsymbol{\chi}\|^2
\end{aligned}
\tag{32}
$$

with $\chi = [\dot{\eta}^{\mathrm{T}} \Delta\omega^{\mathrm{T}} \Delta\Omega]^{\mathrm{T}}$ and $\sigma = \min\{\lambda_{min}(\boldsymbol{D}), k_d, \kappa_d\}$. The analysis above can be summarized as the following proposition:

Theorem 2.
The closed-loop singular system (9), (10) and (17) *is asymptotically stabilized by the output feedback control laws* (31)*; and it is admissible.*

Proof.
Integrating both sides of Eq. (32) yields

$$V(\infty) - V(0) \leqslant -\sigma \int_0^{\infty} \|\chi(t)\|^2 \mathrm{d}t = -\sigma \|\chi\|_2^2 \tag{33}$$

where $\|\cdot\|_2$ is the L_2 norm. Since $V(t) \geqslant 0$, it follows that

$$\sigma \|\chi\|_2^2 \leqslant V(0) - V(\infty) \leqslant V(0)$$

that is

$$\|\chi\|_2 \leqslant \sqrt{V(0)/\sigma}$$

Therefore $\chi(t) \in L_2[0, \infty)$. Furthermore, considering relationship (33) one can infer that $V(t)$ is uniformly bounded in t along the system trajectories, that is

$$\lim_{t\to\infty} V(t) < \infty$$

It implies that $\chi(t)$ is uniformly bounded. Moreover, from system equations it follows that $\dot{\chi}(t)$ is uniformly bounded, thus $\chi(t)$ is uniformly continuous. As $\chi(t)$ is a uniformly continuous function in $L_2[0, \infty)$, from the Barbalat lemma it can be concluded that

$$\lim_{t\to\infty} \chi(t) = 0$$

At that time

$$\varphi \equiv 0 \Rightarrow \dot{\varphi} \equiv 0 \Rightarrow \eta \equiv 0$$

and

$$\Delta\dot{\omega} \equiv \mathbf{0},\ \ \Delta\dot{\boldsymbol{\Omega}} \equiv 0 \Rightarrow \boldsymbol{T} \equiv \mathbf{0},\ \ \tau \equiv 0 \Rightarrow q_v \equiv \mathbf{0},\ \ \Delta\alpha \equiv 0$$

Therefore the solution to the system will eventually converge to

$$\{q_v = \mathbf{0},\ \ \Delta\alpha = 0,\ \ \Delta\omega = \mathbf{0},\ \ \Delta\boldsymbol{\Omega} = 0,\ \ \boldsymbol{\eta} = \mathbf{0},\ \ \boldsymbol{\varphi} = \mathbf{0}\}$$

which implies that the origin of the closed-loop system is asymptotically stable.

Together with Theorem 1, the closed-loop system is admissible according to Definition 2. □

H∞ Optimization

Although the output-feedback-based PD control is able to stabilize singular system (17), spacecraft with flexible structures in practice are affected by many internal and external disturbances, such as spill-over effects, neglected nonlinearities, unmodeled dynamics, parameter variation, and measurement noises in the sensors, which may degrade the closed-loop performance and sometimes lead to instability. Ongoing efforts have been devoted to cope with these uncertainties and improve the performance of the control systems, such as the analysis and synthesis methods proposed in recent works.[17], [18],[19] and [20] Through exemplification, Ref.[21] has emphasized the application of robust control concepts from an integrated control-structure point of view in the overall attitude and orbit control system design.

For the sake of brevity without loss of generality, robust H_∞ optimal control is considered herein against structural parameter variations due to solar array rotation and random noises in angular velocity measurement. Notice that by choosing a solar array angle α_1, the transfer function matrix of the admissible singular system can be obtained by

$$G(s) = C(sM(\alpha_1) - \widehat{A})^{-1}B$$

where s is the Laplace variable. This is true because $\det(sM - \widehat{A})$ exists at any time, as claimed in the proof of Theorem 1. Therefore, based on the robust control theory, an H_∞ optimal control system is constructed as shown in Fig. 2, in which $\boldsymbol{W}_1(s)$, $\boldsymbol{W}_2(s)$,$\boldsymbol{W}_3(s)$ are the weighting functions, $\boldsymbol{r}(t)$ is input, $\boldsymbol{n}(t)$ is the uniform random noise, and$\boldsymbol{G}(s)$ is the generalized control plant with structural uncertainties.

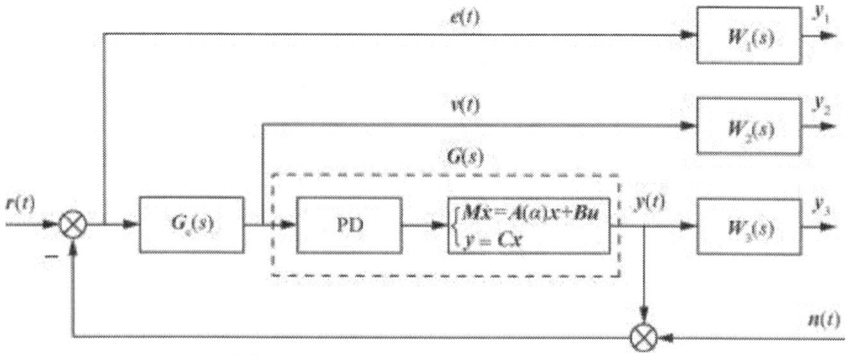

Figure. 2. Schematic of the H_∞ optimal control system.

In such settings, $\boldsymbol{G}_c(s)$ is the resulting H_∞ optimal controller given by the robust synthesis toolbox in MATLAB, where the stability of the whole system is guaranteed by the small-gain theorem. [22] Comparisons and performance analysis of the H_∞ optimal control versus the PD control will be provided by computer simulations in the next section.

SIMULATION EXPERIMENTS

Published in Ref.[11], the finite element analysis data of the American geosynchronous meteorological satellite GOES-8 provide a set of complex flexible spacecraft parameters of system (17) for numerical simulation. Besides a rotating flexible solar array, GOES-8 carries another flexible but fixed appendage (the solar sail), which will be neglected because modal identities as the premise of the mathematical singularity involve properties of the solar array only. Some important parameters of GOES-8 are transcribed as follows for the sake of clarity.

The inertia matrices of the central body and the solar array are given by

$$J_b = \begin{bmatrix} 1028.8716 & -10.9150 & -23.6426 \\ -10.9150 & 1026.2444 & -17.8563 \\ -23.6426 & -17.8563 & 1117.2465 \end{bmatrix} \text{kg} \cdot \text{m}^2$$

and

$$I_s = \begin{bmatrix} 1104.6067 & 9.3024 & 0.0576 \\ 9.3024 & 28.7784 & 2.2498 \\ 0.0576 & 2.2498 & 1076.3055 \end{bmatrix} \text{kg} \cdot \text{m}^2$$

The flexible coupling matrices of the solar array are

$$H_s = \begin{bmatrix} -0.001312 & 33.233892 & 0.002936 & -0.326503 & -0.006997 \\ -0.009509 & 0.329167 & -0.067510 & 4.868742 & -0.116229 \\ -31.935654 & 0.001202 & -5.618361 & 0.128195 & -3.833407 \end{bmatrix} \sqrt{\text{kg}} \cdot \text{m}$$

and

$$P_s = \begin{bmatrix} 6.362845 & -0.000428 & 3.745014 & -0.139602 & 3.858466 \\ -0.060110 & 0.061114 & -0.346686 & -0.007505 & 0.220996 \\ -0.000268 & 7.519367 & 0.001814 & -0.196472 & -0.000649 \end{bmatrix} \sqrt{\text{kg}}$$

Both of them consist of the influence of five flexible modes, namely, the first, third, and fifth modes that are mainly out-of-plane bending, the second mode that is mainly in-plane bending, and the fourth mode that is mainly torsion. Natural frequencies of these modes are $\boldsymbol{\Lambda} = 2\pi$ diag(0.172617, 0.692549, 1.072639, 1.832541, 2.507761) rad/s, and the stiffness matrix is $\boldsymbol{K} = \boldsymbol{\Lambda}^2$. Supposing damping ratios are$\boldsymbol{\xi}$ = diag(0.004, 0.005, 0.0064, 0.008, 0.0085), the damping matrix can be generated by$\boldsymbol{D} = 2\boldsymbol{\xi}\boldsymbol{\Lambda}$.

The two position vectors about the solar array are $\boldsymbol{b} = [-0.02764\ 0.78871\ -0.00844]^T$ m, and $\boldsymbol{r}_s = [-0.02045\ 2.91590\ -0.02455]^T$ m, where $\boldsymbol{r}_s$ is the position vector from the hinge point to the mass center of the solar array w.r.t. F_s (see Fig. 1). With the mass m_s = 85.74 kg, the first moment of inertia of the solar array can be calculated by $\boldsymbol{c}_s = m_s\boldsymbol{r}_s$. Using m_s,$\boldsymbol{c}_s$, and $\boldsymbol{I}_s$, one can investigate the inertial completeness of the flexible coupling matrices by modal identities (12).

For simplicity, the initial attitude error $\boldsymbol{q}(0)$ and rotation angle error $\Delta\alpha(0)$ are presumed to be $[1\ 0\ 0\ 0]^T$ and zero respectively. Suppose that the orbital angular velocity is $\boldsymbol{\Omega}_o$ = 0.060 (°)/s, then the desired angular velocities are given by $\boldsymbol{\omega}_d = [0\ -\boldsymbol{\Omega}_o\ 0]^T$, and $\boldsymbol{\Omega}_d = \boldsymbol{\Omega}_o$. The parameters of the controllers in Eq. (31) are chosen as k_p = 250, κ_p = 25 and k_d = 2500,κ_d = 250. The weighting functions of the H_∞ control are

$$W_1(s) = W_3(s) = \mathrm{diag}\left(\frac{100}{s+0.5}, \frac{100}{s+0.5}, \frac{100}{s+0.5}, \frac{100}{s+1}\right)$$

and

$$W_2(s) = 100 \times \frac{s+1}{s+2} \times \mathrm{diag}(1,\ 1,\ 1,\ 1)$$

where $\boldsymbol{W}_2(s)$ is required to be a proper transfer function matrix to guarantee the existence of the solution to H_∞ optimization. Together with the system transfer matrix$\boldsymbol{G}(s)$ (by setting $\alpha_1 = 0$), the H_∞ optimal controller $\boldsymbol{G}_c(s)$ can be generated by the MATLAB command hinfopt() immediately. The amplitude of the uniform random noise is 1×10^{-5}. For fairness in comparison with H_∞ optimal control, the following low-pass filter is introduced for the PD control to suppress the high frequency noises:

$$F(s) = \frac{1}{1.5s+1}$$

where the cutoff frequency is 2/3 Hz.

With an initial condition (notice the initial vibrations) of

$$x(0) = [\Delta\boldsymbol{\omega}^{\mathrm{T}}(0)\ \dot{\boldsymbol{\eta}}^{\mathrm{T}}(0)\ \boldsymbol{\eta}^{\mathrm{T}}(0)\ \Delta\boldsymbol{\Omega}(0)] = [0\ \ 0\ \ 0\ \ \underbrace{0.001\ \cdots\ 0.001}_{10}\ \ 0]^{\mathrm{T}}$$

simulations of the singular dynamic system (2) and its kinematic equations (9) and (10)under the control of PD and H_∞ optimal are powered by ode45 or the stiff equation solvers in MATLAB/Simulink for 60 s. The time histories of the system's responses and control inputs are shown in Fig. 3, Fig. 4, Fig. 5, Fig. 6 and Fig. 7.

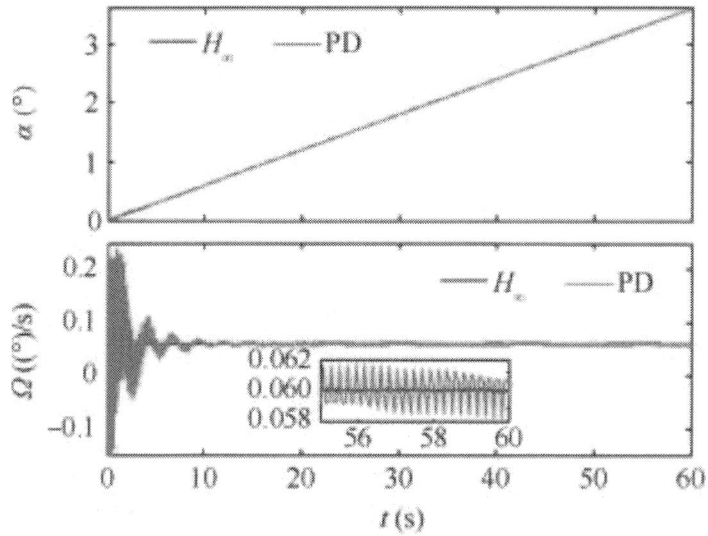

Figure. 3. Solar array rotation angle and velocity.

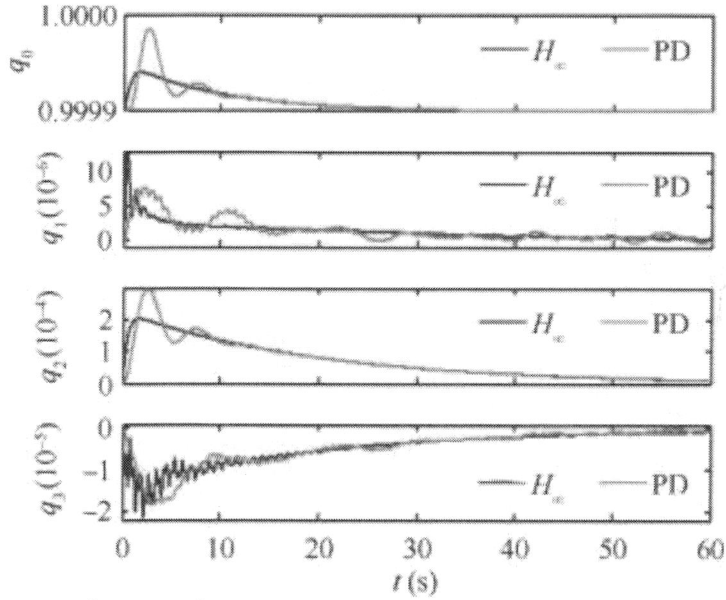

Figure. 4. Satellite attitude errors.

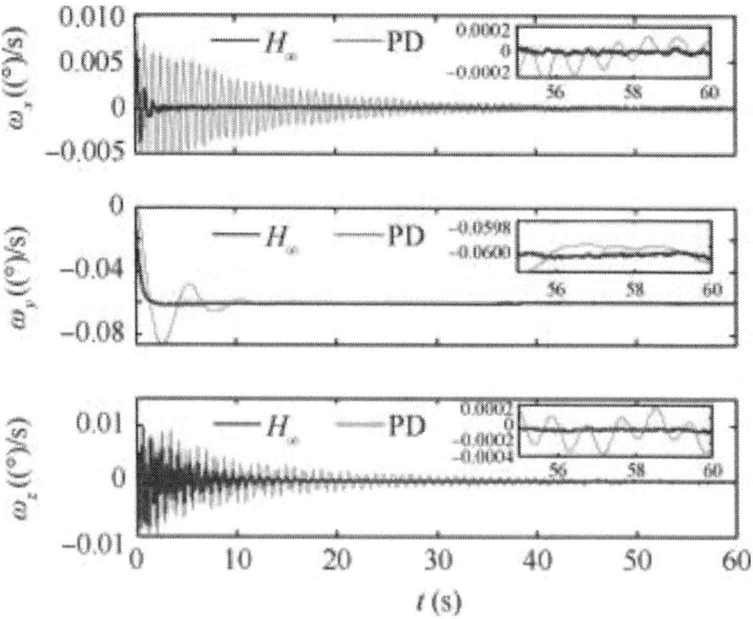

Figure. 5. Satellite angular velocities.

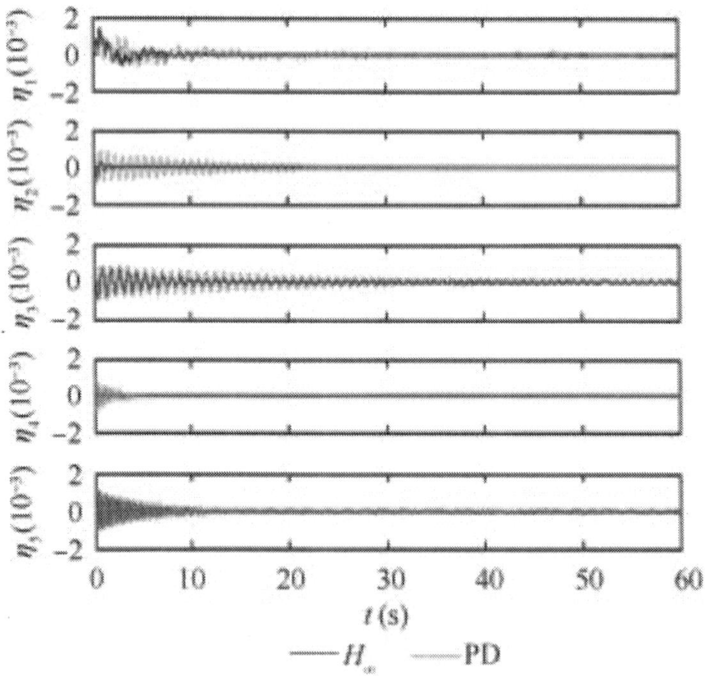

Figure. 6. Solar array flexible modal coordinates.

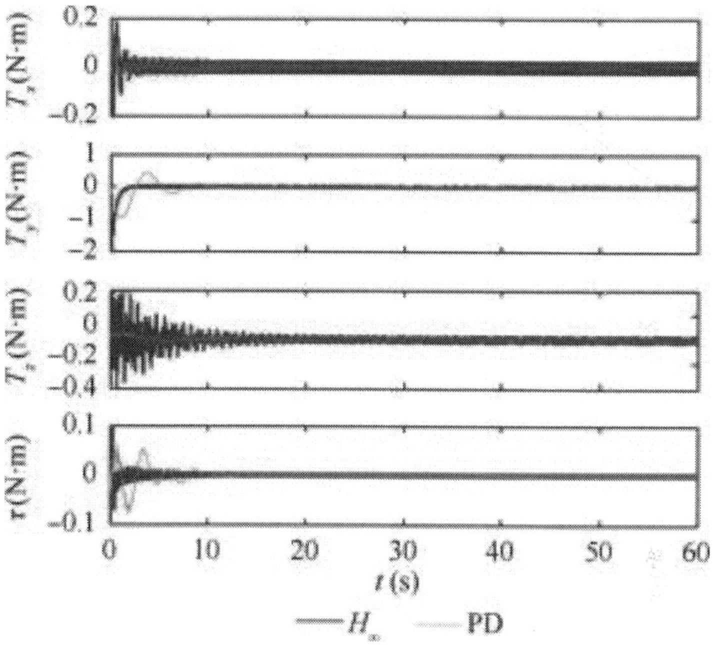

Figure. 7. Control torques.

Clearly, there exists no infinite impulse phenomenon. In the presence of uncertainties in model parameters and measurement noises in the state variables, and both of the control laws are able to drive the core body and the appendage to the prescribed angular velocities without attitude or rotation errors, as illustrated in Fig. 3, Fig. 4 and Fig. 5. However, the H_∞ optimal control reaches the steady state within much less settling time and with less oscillations both in velocities and flexible modal coordinates when compared with the PD plus low-pass filter control. This is because the convergence rates of the flexible modal oscillations in the PD control system depend simply upon the appendage structural damping ($\boldsymbol{D}$), which usually is very weak, whereas the H_∞ optimal control, roughly speaking, considerably improves the performance by extracting the modal information from the velocity feedback and producing the feedforward counteraction to the oscillations. Yet the main drawback of the H_∞ optimal control is that it requires larger input magnitude and higher actuation frequency than the PD control scheme (see Fig. 7), which may constrain its implementation in engineering practice.

CONCLUSIONS

This paper discusses the attitude control of a spacecraft with a rotating flexible solar array. The mathematical singularity of the attitude dynamics is investigated, and a formally singular model is established based on the linearization assumptions and the modal identities, which has not been reported in existing literature. In the framework of singular control theory, an admissible control system is developed and optimized in the sense of H_∞ norm to improve the closed-loop robustness, and then the system is verified by numerical simulation. As a complicated electro-mechanical actuator, however, the dynamics of the solar array drive assembly has not been considered in the spacecraft model. Furthermore, initial instantaneous jumps must have occurred in this complex singular system. The physical essence and detrimental effects of these jumps, as well as their control, call for further research.

ACKNOWLEDGEMENTS

The authors express their thanks to Chen Shoulei and Wang Jiwen for their insightful suggestions.

REFERENCES

1. Ballois SL, Duc G. H1 control of an earth observation satellite. J Guid Control Dyn 1996;19(3):628–35.
2. Valentin-Charbonnel C, Duc G, Ballois SL. Low-order robust attitude control of an earth observation satellite. Control Eng Pract 1999;7(4):493–506.
3. Chre´tien JP, Reboulet C, Rodrigo P, Maurette M. Attitude control of a satellite with a rotating solar array. J Guid 1982;5(6):589–96.
4. Kida T, Yamaguchi I, Chida Y, Sekiguchi T. On-orbit robust control experiment of flexible spacecraft ETS-VI. J Guid Control Dyn 1997;20(5):865–72.
5. Chida Y, Soga H, Yamaguchi Y, Kida T, Yamaguchi I, Sekiguchi T. On-orbit attitude control experiments using ETS-VI by H1 control and two-degree-of-freedom control. Control Eng Pract 1998;6(9):1109–16.
6. Nagashio T, Kida T, Ohtani T, Hamada Y. Design and implementation of robust symmetric attitude controller for ETS-VI spacecraft. Control Eng Pract 2010;18(12):1440–51.
7. Hamada Y, Ohtani T, Kida T, Nagashio T. Synthesis of a linearly interpolated gain scheduling controller for large flexible spacecraft ETS-VIII. Control Eng Pract 2011;19(6):611–25.

8. Nagashio T, Kida T, Ohtani T, Hamada Y. Design and flight test results of robust attitude controller for the ets-vi spacecraft. In: The 18th IFAC World Congress; 2011. p. 5154–9.
9. Xu SY, Lam J. Robust control and filtering of singular systems. Berlin Heidelberg: Springer-Verlag; 2006, p. 12–3.
10. Hablani HB. Modal identities for multibody elastic spacecraft. J Guid 1991;14(2):294–303.
11. Tadikonda SSK. Articulated, flexible multibody dynamics modeling: geostationary operational environmental satellite-8 case study. J Guid Control Dyn 1997;20(2):276–83.
12. Hughes PC, Skelton RE. Modal truncation for flexible spacecraft. J Guid Control 1981;4(3):291–7.
13. Di Gennaro S. Active vibration suppression in flexible spacecraft attitude tracking. J Guid Control Dyn 1998;21(3):400–8.
14. Takaba K. Robust H2 control of descriptor system with timevarying uncertainty. Int J Control 1998;71(4):559–79.
15. Zhang LQ, Huang B. Robust model predictive control of singular systems. IEEE Trans Autom Control 2004;49(6):1000–6.
16. Wang S, Wu M, Jia Z. Matrix inequalities. Beijing: Science Press; 2006, p. 96–7 [Chinese].
17. Wang W, Menon PP, Bates DG, Bennani S. Robustness analysis of attitude and orbit control systems for flexible satellites. IET Control Theory Appl 2010;4(12):2958–70.
18. Gasbarri P, Monti R, Campolo G, Toglia C. Control-oriented modelization of a satellite with large flexible appendages and use of worst-case analysis to verify robustness to model uncertainties of attitude control. Acta Astronaut 2012;81(1):214–26.
19. Gasbarri P, Monti R, De Angelis C, Sabatini M. Effects of uncertainties and flexible dynamics contributions on the control of a spacecraft full-coupled model. Acta Astronaut 2014;94(1): 515–26.
20. Bennani S, Ankersen F, Arcioni M, Casasco M, Massotti L, Silvestrin P. Robust attitude control design for the BIOMASS satellite (earth explorer core mission candidate). In: The 18th IFAC World Congress; 2011. p. 5130–5.
21. Liu M, Xu S, Han C. A backstepping simple adaptive control application to flexible space structures. Chin J Aeronaut 2012;25(3):446–52.
22. Gu DW, Petkov PH, Konstantinov MM. Robust control design with MATLAB. London: Springer; 2005, p. 25–8

CITATION

Dongning Lu, Yiwu Liu, Singular formalism and admissible control of spacecraft with rotating flexible solar array, Chinese Journal of Aeronautics, Volume 27, Issue 1, February 2014, Pages 136-144, ISSN 1000-9361, http://dx.doi.org/10.1016/j.cja.2013.12.010.

Chapter 8

Acoustics and Vibro-Acoustics Applied in Space Industry

Rogério Pirk[1], Carlos d'Andrade Souto[1], Gustavo Paulinelli Guimarães[1] and Luiz Carlos Sandoval Góes[2]

[1] Institute of Aeronautics and Space (IAE) and Technological Institute of Aeronautics (ITA), São José dos Campos, Brazil
[2] Technological Institute of Aeronautics (ITA), São José dos Campos, Brazil

INTRODUCTION

During flight, Expandable Launch Vehicles (ELV) are excited by severe acoustic loads in three phases of flight: lift off, transonic flight and maximum dynamic pressure instant [1]. As such, principles to make onboard equipment compatible with the mission environments must be adopted. At lift off, the highly intense acoustic loads occur; and these levels are usually adopted to qualify payloads and equipments. However, during the transonic flight and maximum dynamic pressure phase, acoustic excitation is also present and such characteristics are also significant for performance evaluation as well as for specific system dynamic qualification/acceptance programs. In this way, noise control treatments (NCT) shall be adopted to alleviate internal vibro-acoustic environments, in view of decreasing costs and developments.

The hostile in-flight environments can damage sensors/conditioners as well as make measurements unreliable. In this way, installation adapters must be designed to protect the sensors. The acoustics of such protective cavities influence the measured sound pressure level (SPL) As such, the cavities must be analyzed and their amplitude-frequency characteristics evaluated. Finally, the measurement corrections, necessary to obtain the actual external SPL, are determined.

Concerning the internal environment found during flights, important launcher subsystems as payload fairing (PLF) and equipment bays shall be investigated and vibro-acoustic analysis can be done, as pointed by [2], [3] and [4]. The PLF is the structural compartment of a launcher where the payload is placed during the flight mission. PLF inner acoustics and its attenuation designs, using virtual prototypes are analyzed using deterministic and statistic techniques. However, when in-flight loading are not characterized, the accounted external air-borne excitation can be those described in [5]. In a similar way, SPL along the launcher structure at lift-off can also be estimated [6]. Furthermore, an alternative procedure to characterize external SPL during flight can also be adopted as described by [7]. Passive vibration control techniques can be used to attenuate structure-borne vibration and the use of viscoelastic materials adding structural damping to reduce the magnitude of vibrations is a well-known solution, usually applied in space and aeronautical industries. On the other hand, the use of active vibration control (AVC) is still considered difficult to be implemented in space industry.

For acoustic noise attenuation, the standard practice is to use passive techniques like blankets ([8] and [9]), which attenuate sound by trapping the energy in the blanket material and dissipating it as heat [10] and Helmholtz resonators tuned to absorb acoustic energy at one or some specific frequencies, typically the cavity frequencies as done by [11].

Another acoustic crucial subject in space industry is combustion instability, since it can severely impair the operation of Liquid Propelled Rocket Engine (LPRE) [12]. In this way, solutions for instability problems in combustion chambers of LPRE as well as solid rocket motors (SRM) are of large interest. In [13], it is described that combustion instability can be verified when the power spectrum of the acoustic pressure measured during tests is analyzed. When an oscillation is observed, i.e., combustion instability, well-defined sound pressure peaks, summed to the background noise are present. Such peaks are correlated to the resonance frequencies of the combustion chamber. In this way, the coupling of acoustic natural frequencies and burning oscillations of the combustion chamber occurs, which can cause instabilities and consequent unexpected behavior as efficiency loss or even the explosion of the engine. In the early developing phases of liquid rocket engines, it is usually proposed the investigation of different combustion chamber configurations [14]. This is usually done in two steps as follows: using theoretical calculation and through experimental measurements. In this way, theoretical and experimental natural frequencies of the acoustic cavity are obtained. Further studies must be performed, applying devices and techniques to attenuate pressure oscillations inside combustion chambers and devices as Helmholtz Resonators, baffles and ¼ wave filters are largely used [15] and [16].

This chapter describes three case studies applied on space industry. Firstly, analytical and numerical modelings of in-flight external microphone protection devices are described. Testing procedures and the SPL measurement correction factors determination are also presented. As a second case study, deterministic and statistical coupled vibro-acoustic analysis techniques are used to estimate PLF internal SPL at lift-off as well as to assess the effect of including NCT (blanket materials) on its skin. The modeling procedures and experimental ground test are described. Finally, in the third case, the acoustic characterization of combustion chambers is presented. Cold tests are described as well as the theoretical modeling procedures. The pressure attenuation control technique using Helmholtz Resonators are also presented. In all three case studies, theoretical x experimental results are depicted.

EXTERNAL ON BOARD MICROPHONES INSTALLATION DEVICES

At lift off, the source of the acoustic noise is the gas stream ejected by the motors (Fig. 1). Such acoustic pressure lies in the range of 140 to 180 dB near the rocket and is very close to an acoustic diffuse field (ADF) noise. At transonic flight the launcher is excited by the turbulent boundary layer (TBL) in the neighborhood of the shock waves. According to [17], when the maximum dynamic pressure occurs, the unsteady pressure field applied to the launcher is due to aerodynamic noise. The characteristics of such noise are very different from those at lift off. Non-attached flows increase the pressure in low-frequencies, which excite the launcher first structural modes. A simulation of the VLS-1 flight aerodynamics was done by Academician V. P. Makeyev State Rocket Centre (SRC- Makeyev), as shown in Fig. 2. Notice that the upper nose and 1st stage noses are the most exposed regions to aero-acoustic noise.

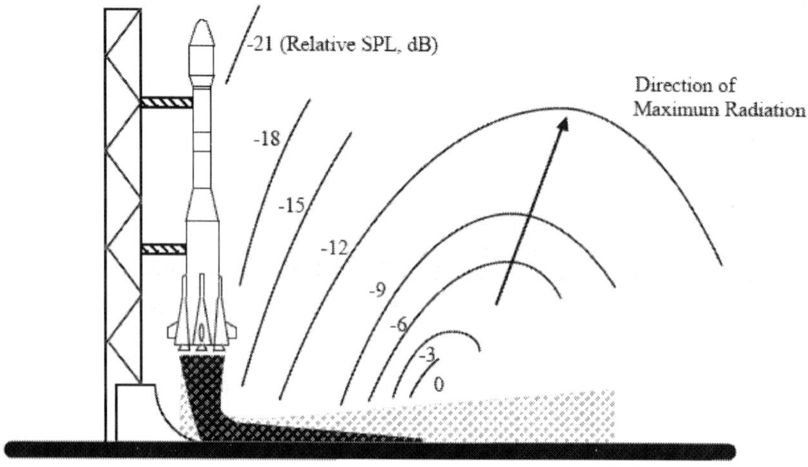

Figure 1. Acoustic noise at lift off

In view of having a good characterization of the in-flight acoustic loads acting upon the launcher structure, external acoustic measurements are required. Due to the high SPL and hostile environments found during flight, special microphones and adapters are specified. Such adapters must be designed in order to provide appropriate microphone/pre-amplifier installation and protection. Besides, when necessary, measurement correction procedures must be adopted. In this way, measurement programs for in-flight external acoustic characterization shall be developed, which may take into account three main phases as: preparation for experimental studies and acoustic testing sensors, ground development testing of acoustic sensors and methodology for reading acoustic pressures during flight.

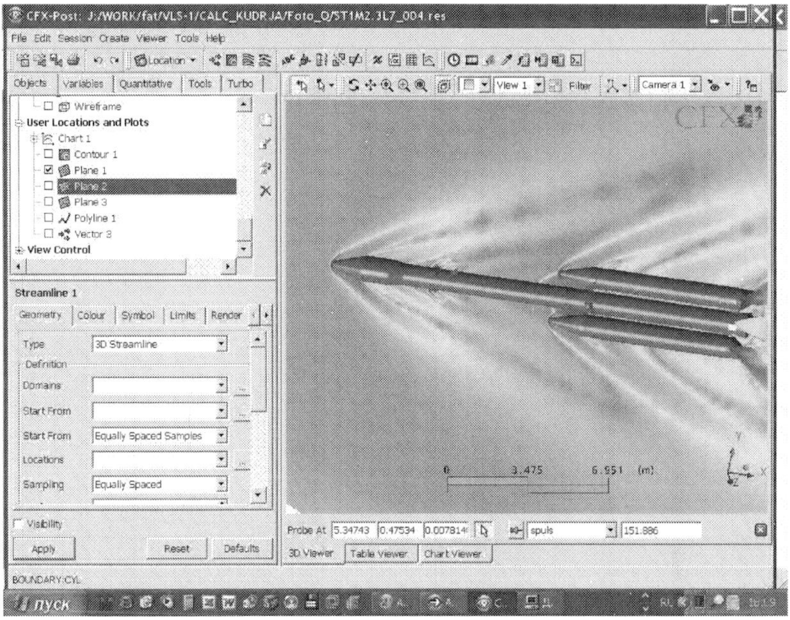

Figure 2. In-flight aero-acoustic noise

Three different adapters for ¼'' microphones were conceived, as described by [18]. On the upper parts of the launcher, two different configurations can be adopted. Firstly, at the PLF, heating and propellant dust effects are not significant; therefore, a structure flush installation (Fig. 3a) can be used. In this case, the measured SPL can be read directly. The second configuration, straight adapter (Fig. 3b), is applied for microphones installed near the equipment bays, where one has temperature and dust influences and, therefore, the sensor/conditioner must be protected. For such an assemblage, the protection channel dynamics directly affect the sensor response and, as a result, a measurement correction must be done.

On the bottom, the intense SPL at lift off generate a severe acoustic excitation of the first stage back modules region. Highly hostile dust, hot gas flow, heat flux and temperature environments are present during the motors operations. Nevertheless, the angular adapter must be used to install acoustic microphone/pre-amplifier, as shown in Fig. 3c. Notice in Figs. 3b and 3c, that the adapters were designed with small acoustic straight and angular cavities, respectively. When acoustically excited, the acoustical responses of such cavities directly influence the measured SPL, since the external pressure excitation profile and the measured signal are related by the cavity transfer function.

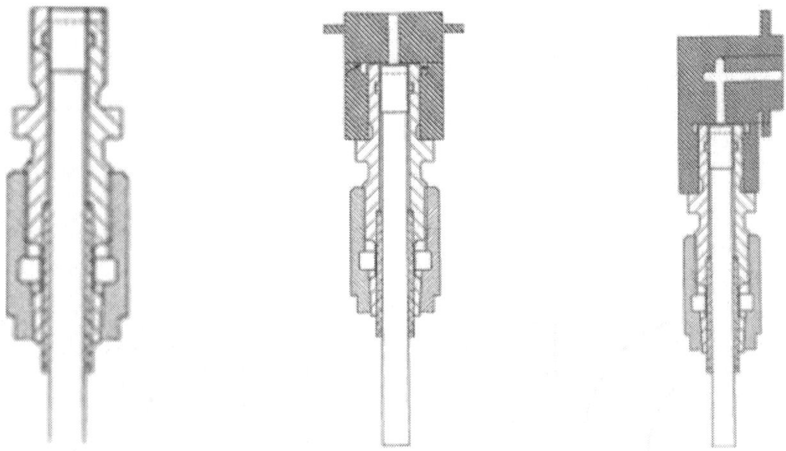

(a (b) (c)

Figure 3. a) Flush adapter; (b) Straight adapter; (c) Angular adapter

In order to determine straight and angular adapters' dynamics, analytical and numerical calculations are done. The transfer functions of these channels are evaluated during ground acoustic tests, on which acoustic excitation with SPL close to those expected during flight is used to excite the cavities. Consequently, the measured SPL as well as the channels transfer functions are determined. Finally, the measurement corrections are determined, which may be applied when these adapters are used.

Mathematical Models

Analytical Model

In view of describing the dynamical behavior of the protective channels, one can assume the straight and angular channels as Helmholtz Resonators, which the channels and the space for microphone installation are accounted as the resonator throat and volume, as described by Eq. (1) [19].

$$f_0 = \frac{c_0}{2\pi}\sqrt{\frac{S}{V_0(l+l_c)}} \quad (1)$$

where: f_0: natural frequency, c_0: sound speed, S: cross section of the resonator throat, V0: volume of the resonator cavity, l: length of the resonator throat, l_c=0.8r: end correction and r: radius of the resonator throat cross section.

By considering the dimensions of the adapters into Eq. (1), one can calculate the natural frequencies of the straight and angular channels shown in Figs. 3b and 3c. These calculations are the starting point to assess the accuracy of the numerical models, built by using the Finite Element Method, once analytical x numerical natural frequencies can be compared.

Numerical Model by Finite Element Method (Fem)

In a similar way as in structural dynamics, an acoustic cavity FEM model will have an acoustic stiffness matrix $[K_a]$, an acoustic mass matrix $[M_a]$, acoustic excitation vectors $\{F_{ai}\}$ and an acoustic damping matrix $[C_a]$. The combination of these components yields the acoustic finite element model, which can be solved for the unknown nodal pressure values p_i[20].

$$\left([K_a]+j\omega[C_a]-\omega^2[M_a]\right)\cdot\{p_i\}=\{F_{ai}\} \quad (2)$$

Acoustic finite element models of the three adapters cavities are built. All cavities' surfaces were considered as rigid walls but the openings that are in direct contact with the external acoustic environment. In such cases, opened surfaces approximated using prescribed nodal pressures (equal to 0 for the eigenanalysis) were considered. The fluid inside the cavities is assumed as air at 15° C (c=340 m/s, ρ=1.225 Kg/m^3, values used in this entire chapter).

Linear tetrahedral fluid elements are used in all three meshes. In order to have good prediction accuracy in the frequency range of interest, the general rule of thumb that requires at least 6 elements per wavelength is adopted. The main meshing characteristics are described in Table 1.

The acoustic load generated at lift off is simulated as an acoustic diffuse field (ADF). According to [21], an ADF is defined as an acoustic field in which the SPL is equal at any location and have an identical energy distribution in all directions. Such ADF can be obtained in an acoustic reverberant chamber, where the reflections along the rigid walls lead to this field. A formal way to describe an ADF consists on superimposing an infinite number of uncorrelated plane waves through different directions. In a FEM model, a finite number of uncorrelated plane waves can be generated and the pressure due the superposition of all the uncorrelated plane waves can then be applied as prescribed nodal pressures on the cavity's open surface (see [22]).

Table 1. Meshes data

	Mesh		
Parameters	Flush adapter	Straight adapter	Angular adapter
Number of elements	4,419	3,986	10,266
Number of nodes	1,009	980	2,367
Maximum frequency (6 elements/wavelength)	45,374 Hz	44,593 Hz	44,199 Hz

The FEM models of the three adapters are shown in Figs. 4a, 4b and 4c. The cavity's transfer functions are calculated by imposing prescribed nodal pressures in the nodes marked with small green arrows. In order to save time and computational efforts, a modal solution method is adopted using the first 14 modes. A modal damping of 5% is considered in these calculations.

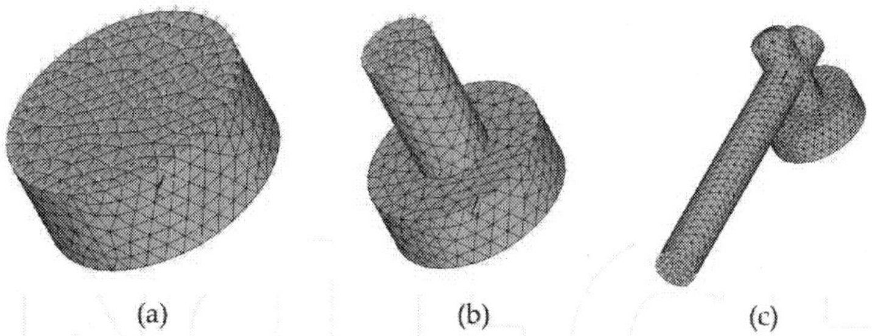

Figure 4. a) Flush adapter mesh; b) Straight adapter mesh; c) Angular adapter mesh

Experimental Set Up

An experimental unit is conceived to characterize all three adapters, as shown in Fig. 5. The experimental unit is placed into an acoustic reverberant chamber and submitted to an ADF, with a frequency profile shown in Fig. 6, which impinges the plate where the adapters and microphones/conditioners are installed. Care is taken to assure that the plate has structural response similar to that found along the launcher skin. Accelerometers are installed on the plate to measure the acoustically induced structural vibration.

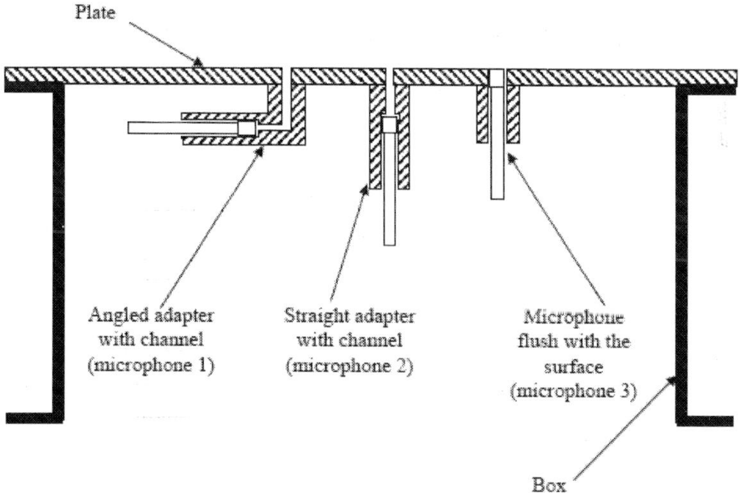

Figure 5. Experimental unit

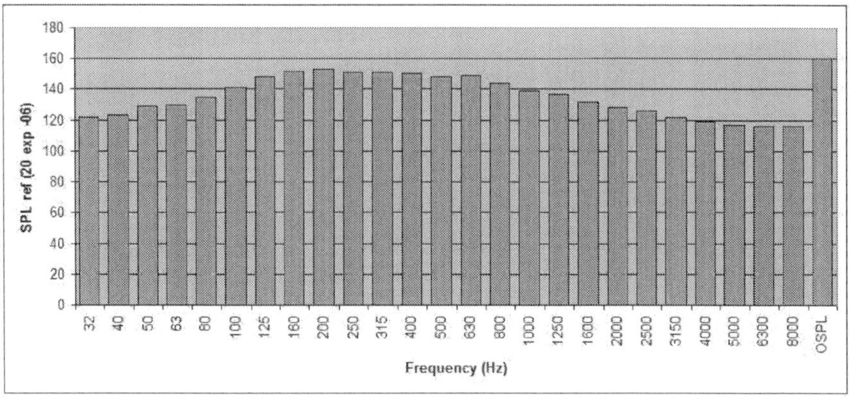

Figure 6. Excitation profile Overall Sound Pressure Level (OSPL) – 160 dB

Methodology for the External Measured SPL Correction

Characterization of the Adapters´ Cavities

The amplitude-frequency characteristics of such adapters must be accounted to determine the acoustic levels measured during the flight mission, since the device channels operate like filters. The SPL measured by the microphone installed with the flush adapter (Fig. 4a) is considered as the reference. Then the adapters transfer functions relating the input signal given by the microphone with the flush adapter and the output signal given by the microphones with the protective adapters (angular or straight) can be obtained by Eqs. (3) and (4)

$$M1(\omega) = H_{M1}(\omega)M3(\omega) \tag{3}$$

$$M2(\omega) = H_{M2}(\omega)M3(\omega) \tag{4}$$

where: $H_{M1}(\omega)$and$H_{M2}(\omega)$: angular and straight adapter transfer functions, respectively.

*M*1, *M*2, *M*3: measured SPL into angular, straight and flush adapters.

Results

Analytical and numerical natural frequencies results are compared with those obtained by acoustic testing in Table 2. As pointed out before, the measured SPL with the flush adapter can be read directly. In this way, only the adapters shown in Figures 4 b and 4c are considered.

Table 2. Theoretical x Experimental natural frequencies (in Hz)

	Straight Adapter				**Angular Adapter**			
Mode	Analytical	FEM	Testing	FEM-Testing Difference (%)	Analytical	FEM	Testing	FEM-Testing Difference (%)
1	5,000	5,396.60	5,202	3.74	2,300	2,204.70	1,953	12.88
2	-	27,032	-	-	-	8,660.30	7,831	10.58

Disagreements between test and predicted resonance frequencies can be explained by possible inaccuracy in microphone installation and complicated shape of the angular channel. The characterization of the adapters´ cavities is performed numerically, by calculating $H_{M1}(\omega)$ and $H_{M2}(\omega)$ (Eqs. (3) And (4)) using FEM. The calculated response functions of the angular and the straight adapter models are compared to those obtained experimentally in Fig. 7 (experimental frequency resolution Δf = 18.78 Hz; numerical frequency resolution Δf = 20.00 Hz).

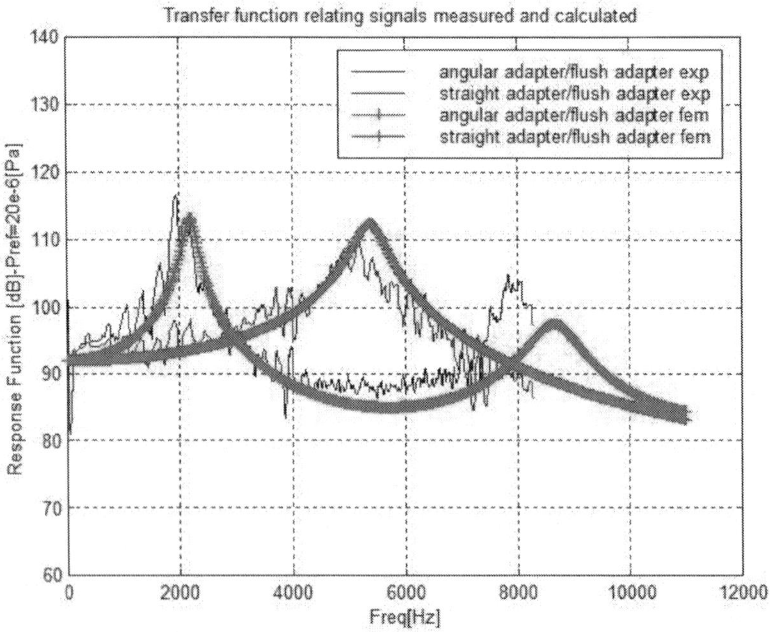

Figure 7. Microphones with adapters transfer functions (numerical and experimental)

The theoretical and experimental response functions show good agreement, with a shift in the second resonance peak for the microphone with angular adapter. The errors can be caused by: bad characterization of the ADF spectral distribution into the FEM model; adapter's geometry complexity; microphone installation inaccuracy or a combination of some of these factors.

Equations 3 and 4 show that the external noise (given by M3 (ω)) can be identified by knowing the internal noise and the inverses of the transfer functions for the angular and straight adapters.

VIBRO-ACOUSTIC MODELLING OF PAYLOAD FAIRINGS (PLF)

A complete survey of PLF vibro-acoustic environment must be carried out, in order to determine its inner SPL. In this respect, it is important to have reliable numerical tools that can predict the responses of ELV systems, subjected to in-flight acoustic loads and that enable NCT design.

Low-frequency coupling techniques are used to estimate a PLF dynamic behavior. The fairing body and its inner acoustic domain are analyzed by using Finite Element Method (FEM) and Boundary Element Method (BEM). Structural FEM/fluid FEM and structural FEM/fluid BEM modeling techniques are then applied. In order to simulate the lift off acoustic excitation, an ADF of 145 dB OSPL is applied on the fairing body and coupled calculations are done from 5 to 150 Hz, which yielded the acoustic and skin responses for both models. Modal expansion and semi-modal expansion model techniques are applied, respectively.

For the high-frequency analysis, it is applied the Statistical Energy Analysis (SEA) technique, for a frequency range from 5 to 8,000 Hz. The 145 dB OSPL excitation is applied to the structural panels of the fairing and the acoustic and structural mean responses are calculated.

In view of validating the numerical predictions for the fairing, acoustic test is done to measure the acoustics inside the PLF. The PLF is submitted to 145 dB OSPL in a 1,200 m^3 acoustic reverberant chamber and microphones are positioned in its inner domain.

The implementation of sound absorption blankets is applied as a control technique to attenuate acoustic noise from medium- to high-frequency bands. SEA is a technique for high-frequency analysis; therefore, adequate to assess the influence of blankets on space systems. The generated SEA fluid-structure model is used to calculate internal SPL, with single-, double-, and multi-layered noise control treatments (NCT). Two NCT modeling approaches are used to simulate the effect of blanketing the fairing cavity:

I. acoustic materials Biot´s parameters, given by the manufacturer;
II. material samples absorption coefficient, measured in a Kundt Tube.

Model Description

The analyzed fairing is hammerhead type geometry and is composed of the body structure, functional components as electric and pyrotechnic components of the ejection system, mechanisms as well as the exterior cork liner. Figures 8a and 8b show the Brazilian VLS fairing structure.

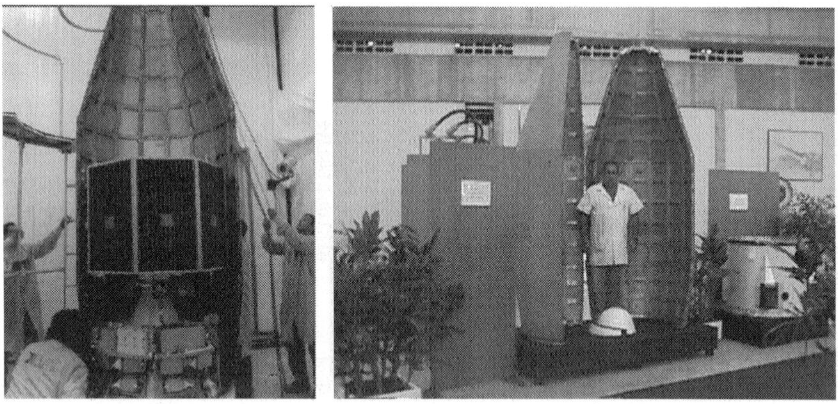

(a) (b)

Figure 8. a) PLF structure; b) PLF structure

Modelling Methodology

Low-Frequency Modeling (Deterministic) Techniques

In view of predicting the operational fairing cavity SPL, both the dynamic displacements of the fairing structure as well as the acoustic pressure fields at the interior and the exterior side of the fairing should be considered. In this study, however, the fluid-structure coupling interaction between the structural displacements and the exterior acoustic pressure field is neglected. The exterior acoustic pressure is assumed to be a known external excitation for the vibro-acoustic system, consisting of the fairing body and the internal acoustic cavity. The FEM and BEM are the most appropriate numerical techniques for the (low-frequency) dynamic analysis of this type of vibro-acoustic system.

FEM based models for coupled vibro-acoustic problems are most commonly described in an Eulerian formulation, in which the fluid is described by a single scalar function, usually the acoustic pressure, while the structural components are described by a displacement vector. The resulting combined FEM/FEM model in the unknown structural displacements and acoustic pressures at the nodes of, respectively, the structural and the acoustic FEM meshes are [20],

$$\left(\begin{bmatrix} K_S & K_C \\ 0 & K_A \end{bmatrix} + j\omega \begin{bmatrix} C_S & 0 \\ 0 & C_A \end{bmatrix} - \omega^2 \begin{bmatrix} M_S & 0 \\ -\rho K_C^T & M_A \end{bmatrix} \right) \begin{Bmatrix} w_i \\ p_i \end{Bmatrix} = \begin{Bmatrix} F_{Si} \\ F_{Ai} \end{Bmatrix} \tag{5}$$

Where:

K_S – Structural stiffness matrix M_S – Structural mass matrix C_S – Structural damping matrix

K_C – Fluid – structure coupling matrix

M_A – Fluid mass matrix K_A – Fluid stiffness matrix $_{CA}$ – Fluid damping matrix

w_i – Structural displacements nodal vector p_i – Acoustic pressures nodal vector

F_{Si} – Force nodal vector F_{Ai} – Acoustic sources nodal vector

In comparison with a purely structural or purely acoustic FEM model, the coupled stiffness and mass matrices (Eq. (5)) are no longer symmetrical due to the fact that the force loading of the fluid on the structure is proportional to the pressure, resulting in a cross-coupling term

KC in the coupled stiffness matrix, while the force loading of the structure on the fluid is proportional to the acceleration, resulting in a cross-coupling term $M_C = -\rho K_C^T$ in the coupled mass matrix.

Low-frequency vibro-acoustic problems can also be modeled by describing the structural behavior in a FEM model and the fluid behavior in a BEM model. In the same way as in the FEM/FEM technique, deterministic FEM/BEM models are usually described by acoustic double layer potential and structural displacement, which are the field variables. Equation (6) presents the resulting combined FEM structural displacements and BEM acoustic pressure differences at the nodes, for a coupled FEM/BEM mesh [23].

$$\begin{pmatrix} K_S + j\omega C_S - \omega^2 M_S & L_C \\ L_C & \dfrac{D}{\rho_0 \omega^2} \end{pmatrix} \begin{Bmatrix} w_i \\ \mu_i \end{Bmatrix} = \begin{Bmatrix} F_{Si} \\ F_{Ai} \end{Bmatrix} \tag{6}$$

Where *LC* is the fluid-structure coupling matrix, *D* is the BEM acoustic matrix of coefficients and *μi* is the nodal vector of double layer potentials. In deterministic models, the dynamic variables within each element are expressed in terms of nodal shape functions, usually based on low-order (polynomial) functions. Since these low-order shape functions can only represent a restricted spatial variation, a large number of elements is needed to accurately represent the oscillatory wave nature of the dynamic response. A general rule of thumb states that for fluid-structure interactions, at least 6 (linear) elements per wavelength are required to get reasonable accuracy. Since wavelengths decrease for increasing frequency, the FEM model sizes, computational efforts and memory requirements increase also with frequency. As a result, the use of FEM and BEM models is practically restricted to low-frequency applications. In comparison with uncoupled structural or acoustic problems, this practical frequency threshold becomes significantly smaller for coupled vibro-acoustic problems, since a structural and an acoustic problem must be solved simultaneously. Moreover, the matrices in a coupled model are no longer symmetrical, so that less efficient non-symmetrical solvers must be used. As a consequence, the computational effort, involved with the use of coupled FEM/FEM and FEM/BEM models for real-life vibro-acoustic engineering problems, becomes large at very low frequencies.

In order to obtain coupled vibro-acoustic response predictions within reasonable computational efforts, the dimensions of the FEM/FEM problem (Eq. (5)) have to be reduced. The most applied technique for model reduction is the modal superposition technique, which expresses the unknowns of the system in terms of a modal basis, resulting in a set of unknown modal participation factors, whose size is much smaller than the size of the original set of unknowns. A modal expansion in terms of uncoupled structural and acoustic modes is performed by using computationally efficient symmetric eigenvalue algorithms and requires much less computational effort than the use of vibro-acoustic (coupled) modes. However, a large number of high-order uncoupled acoustic modes is required to accurately represent the normal displacement continuity along the fluid-structure interface.

In a FEM/FEM virtual prototype, a modal expansion in terms of uncoupled structural and uncoupled acoustic modal bases is used, in order to keep the computational efforts within reasonable limits. On one hand, structural wavelengths are much smaller than acoustic wavelengths, so that the structural FEM mesh of the fairing must be finer than the acoustic FEM mesh of the inner cavity. On the other hand, due to the continuity of the normal structural and fluid displacements along the fluid-structure coupling interface, both meshes must be compatible in this region. In this framework, the following modeling methodology is adopted: 1) A fine FEM mesh of the fairing is used for the construction of the uncoupled structural modal data basis. 2) The resulting modes are then projected onto a FEM coarse mesh of the fairing structure. 3) For the acoustic cavity FEM mesh, the same mesh density is used along the fluid-structure coupling interface as the PLF structural coarse mesh, while the mesh density is slightly decreased towards the central axis of the cavity. 4) The uncoupled modes, resulting from this acoustic FEM mesh, together with the projected structural modal basis are used in a coupled FEM/FEM model. It is important to highlight that the coarse structural mesh has only the shells of the fairing structure, while all reinforcing beams are omitted, since it is assumed that these stiffeners have no significant effect on the fluid-structure coupling interaction, while their presence would increase the computational load of the modeling process.

For the case of the FEM/BEM problem (Eq. (6)), the modal expansion cannot be used, since the frequency dependency of the matrix coefficients in the acoustic part prohibits a standard eigenvalue calculation. Such as, the semi-modal approach, which uses only the expansion of the structural modal data basis, is applied. As mentioned above, BEM drawbacks as fully population of the matrices, complex and frequency dependent models result in a coupled FEM/BEM model less efficient than coupled FEM/FEM model. Therefore, the rule of thumb of 6 (linear) elements per wavelength becomes prohibitive for the actual fairing fluid-structure study. Such a way, a coarsest structural mesh may be generated and the same adopted frequency range for FEM/FEM model is kept for this FEM/BEM model, even considering that the structure has not enough discrete density. However, the modal data basis calculated using the structural fine mesh can assure good results for the structural displacements, in this fluid-structure model, since the expansion in terms of such a data basis is used on these Frequency Response Analysis (FRA) computations. The displacement continuity of the structural and acoustic meshes (same density in the fluid-structure interface) is considered to perform the link and calculate the coupled dynamic skin displacements and acoustic cavity pressure responses, for both FEM/FEM and FEM/BEM models.

High-Frequency Modeling (Statistical) Technique

A characteristic of high-frequency analysis is the uncertainty in modal parameters. The resonances and mode shapes show great sensitivity to small variations of geometry, construction and material properties. In addition, programs used to evaluate mode shapes and frequencies are known to be inaccurate for higher modes. In light of these uncertainties, a statistical model of the dynamic parameters seems natural and appropriate. As an alternative method for higher frequency analysis of the inner cavity of fairings, Statistical Energy Analysis (SEA) approach is proposed. This approach is the description of the dynamic system as a member of a statistical population or ensemble, whether or not the temporal behavior is random. SEA emphasizes the aspects of this field dynamical study.

The SEA equations express the energy balance of different subsystems in a model [24]. Some subsystems have direct power input of an independent source, e. g. an excitation force on a structural component, a sound power source in an acoustic medium etc. In general, subsystems can receive power (input power from external sources), dissipate power (internal losses due to damping) and exchange power with other subsystems to which they are coupled (losses due to coupling). SEA fundamental hypothesis as dissipation losses in relation to the energy variable and modal energy proportionality from connected subsystems are used to yield the SEA matrix equation of complex structures. The distribution of the dynamical response in the system due to some excitation is obtained from the distribution of the energy among the mode groups, based on a set of power balance equations for the mode groups.

Fem Structural Meshes

The fairing body is divided in five surfaces. The surfaces are discretized by using 4-noded quadrilateral shell elements, while 2-noded beam elements are used for the circumferential and the axial stiffeners. To account for the mass of the cork blanket on the exterior fairing surface, a distribution of concentrated mass elements are attached to the fairing nodes.

A total of 174 structural modes in a frequency range up to 220 Hz have been identified. Table 3describes the main structural modes calculated by using FEM in the range up to 150 Hz.

Table 3. Fairing body structural modes

mode	frequency (Hz)
first bending	38.637
first breathing	77.821
second breathing	92.694
first longitudinal	108.749
first torsion	125.772
second bending	150.735

Table 3 shows that the first structural bending mode of the fairing is identified at 38.6 Hz, while the second mode is at 150.7 Hz. Figures 9 a and 9b show the referred structural modes.

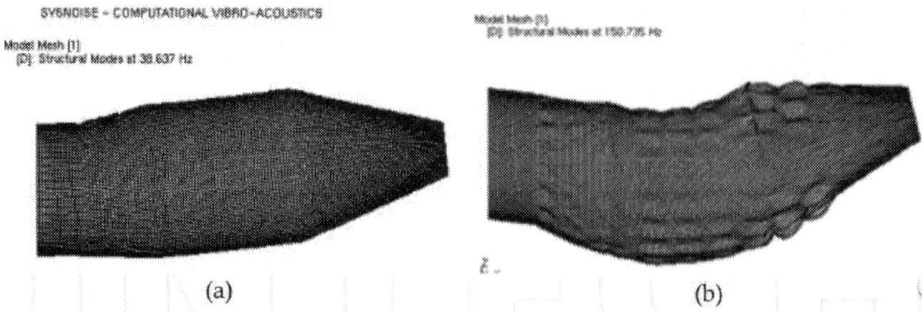

Figure 9. a) First structural bending mode; b) Second structural bending mode

Fem and BEM Acoustic Meshes

The acoustic FEM mesh consists of 119,577 nodes and 110,238 elements (106,050 8-noded hexahedral elements and 4,188 6-noded pentahedral elements). The cavity is considered filled with air at 15° C. The cavity's bottom and top faces are assumed to be acoustically closed (rigid walls). The acoustic mesh generation takes into account the meshes compatibility on the fluid-structure interface.

A total of 80 acoustic modes in a frequency range up to 566 Hz were identified. Acoustic wavelengths are bigger than structural wavelengths. Such that, a large number of high-order uncoupled acoustic modes is required to accurately represent the normal displacement continuity along the fluid-structure interface. That is why higher frequency range is used to describe the acoustic modal behavior of the fairing. Table 4 describes the acoustic modes in the frequency range up to 150 Hz.

Table 4. Fairing cavity acoustic modes

Mode	Frequency (Hz)
rigid body	0
first longitudinal	63.491
second longitudinal	112.129

It can be noticed in table 4 that the first and second acoustic modes of the fairing cavity are identified at 63.5 Hz and 112.1 Hz, respectively. Figures 10a and 10b show the referred acoustic modes.

The BEM acoustic mesh is a 2-D coarsest mesh. Therefore, as the coupled FEM/BEM equation is frequency dependent (Eq. (6)), the acoustic modes are not considered in the acoustic pressure calculations (semi-modal reduction model).

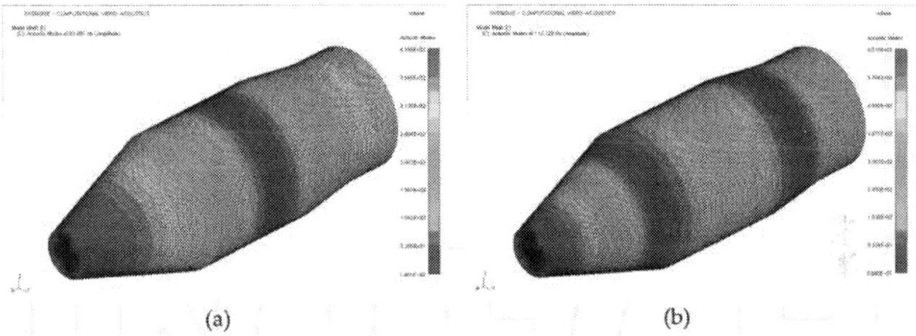

Figure 10. a) First acoustic longitudinal mode; b) Second acoustic longitudinal mode

Model Excitation

A uniform pressure loading is simulated by applying a normal point force varying harmonically on all nodes of the fairing shell elements. The force value is defined such that the total load is equivalent to a uniform pressure of 145 dB OSPL. Link of the acoustic and structural parts is done as well as the structural modal data basis is projected to the coarse and coarsest meshes.

In this way, all the meshes, modal data bases and excitation, needed to perform low-frequency calculations, using coupling fluid-structure techniques are ready. Next step is to perform FRA calculations for both models.

Sea Fairing Vibro-Acoustic Model

The fairing body is divided in four surfaces, as shown in Fig. 11a. To account for the rib-stiffened plates of the surfaces 2, 3 and 4, the SEA structural fairing model considers connected plates and beams (longitudinal and circular). The plate structural subsystems are generated as singly curved shells and uniform plates. Shell surface 1 has a thickness of 3mm and is modeled as a simple plate of aluminum (E=72 GPa, ν=0.29, ρ=2750 kg/m^3), while the other three surfaces are 0.8 mm thick and made of an aluminum alloy (E=72 GPa, ν=0.29, ρ=7000 kg/m^3). The circular and longitudinal beams are modeled by assigning the same material as the shells of the surfaces 2, 3 and 4 (Figure 11a). Damping loss factors of 1% (for flexure, extension and shear propagating waves) are assigned to the plates and beams subsystems, in order to account for the internal loss factors.

A total of 72 beams (44 longitudinal and 28 circular) and 8 shells (02 singly curved shells of the adaptor, 02 singly curved shells of the lower cone, 02 singly curved shells of the main cylinder and 02 singly curved shells of the upper cone) compose the structural SEA model. Figure 11a shows the SEA plates and beams generated to model the VLS-1 fairing structure. The external blanketed treatment of cork on the surfaces 2, 3 and 4, was simulated in this model as material addition. The layered area and the density of the cork were considered to assign this mass.

The acoustic environment inside the fairing was generated by starting from the structural model. This acoustic cavity was created considering air at 15° C as the fluid as well as the dimensional parameters of the fairing. The top and bottom face of the cavity were assumed to be acoustically closed. Figure 11bpresents the 3D acoustic cavity of the fairing.

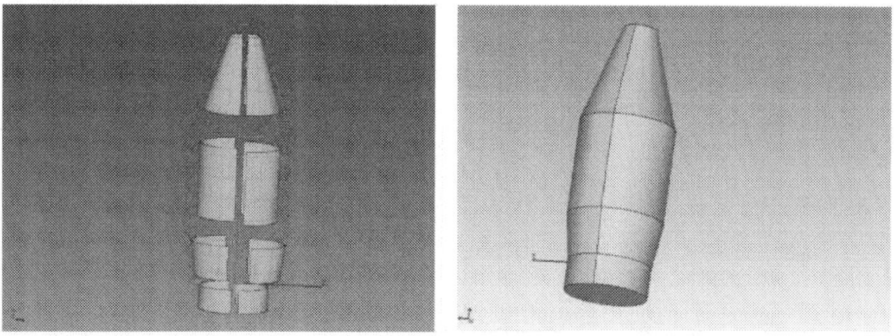

(a) (b)

Figure 11. a) Structural subsystems (shells, circular beams and longitudinal beams) b) Acoustic cavity of the fairing

The coupling boundary between all the structural and acoustic subsystems is modeled to consider the transmission of power across the junctions. A junction is comprised of connections to any number of coincident subsystems. As a result, all the subsystems that share common nodes are connected by point, line and area junctions and all the appropriate wavefields are connected as well as the corresponding coupling loss factors (CLF) between subsystems are created.

The estimated sound pressure levels at the lift off are assigned to the SEA model. Only elements with large surface areas, as plates and panels, are considered to be susceptible to acoustic excitation ([11], [24] and [25]). An ADF of 145 dB OSPL (Fig. 12) is applied to the plates of the SEA fairing model, which simulates the power input into a structural plate or shell element.

Analysis Results

In view of having a complete knowledge of the fairing dynamic vibro-acoustic behavior, the fairing structural skin as well as its inner acoustic domain responses should be presented. However, since this chapter concerns acoustics, the body structural displacements are not presented here. Below, the obtained results of the acoustic behavior applying vibro-acoustic low-frequency and high-frequency analysis techniques predictions are presented.

Low-Frequency Techniques

FEM/FEM Response Calculations

A modal expansion in terms of 174 uncoupled structural and 80 uncoupled acoustic modes is used for the coupled calculations. A modal damping of 1% is assigned to all structural modes. All calculations are performed with a frequency resolution of 1 Hz. Figure 13 shows the low-frequency acoustic pressure spectra of the PLF for the case of a uniform exterior pressure loading, using FEM/FEM coupling analysis. It can be seen that the low-frequency pressure is dominated by the first longitudinal mode around 63.5 Hz and the second longitudinal mode around 112.1 Hz.

FEM/BEM Response Calculations

The same structural modal expansion as used for FEM/FEM is used for this FEM/BEM response calculations. Due to the frequency dependency of the boundary integral equation, the acoustic modal basis can not be used. A damping of 1% is assigned to all structural modes. All calculations are performed with a frequency resolution of 2 Hz. Figure 13 presents a comparison of the computed inner cavity space averaged acoustic pressure using FEM/FEM and FEM/BEM techniques.

High-Frequency Technique

The energy balance (levels and interactions) between different subsystems of the SEA model is calculated. The interest frequency range is 5 to 8,000 Hz, by third octave bandwidth. As mentioned before, SEA technique is more effective in higher frequencies, where dynamic systems present higher modal density. The vibro-acoustic responses of the fairing, using SEA technique, are shown in Fig 13.

It is important to highlight that for the low-frequency range, SEA analysis results are not reliable, since the accuracy of the SEA technique is proportional to the modal density [24].

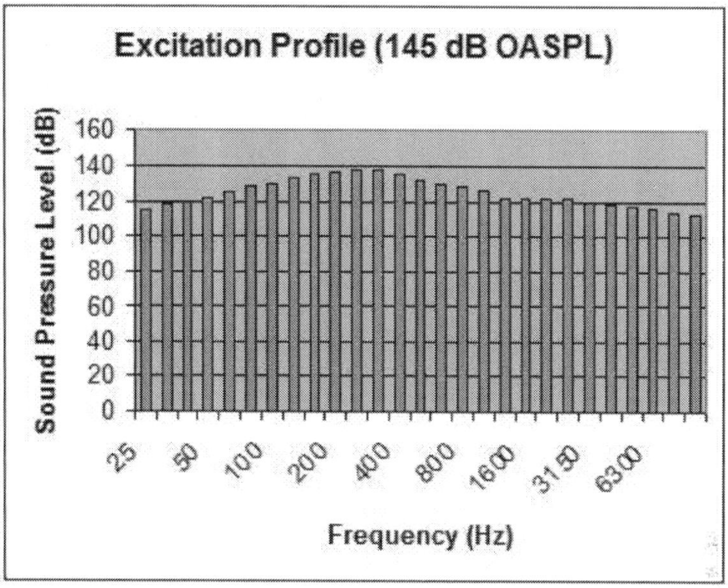

Figure 12. Acoustic diffuse field at lift off

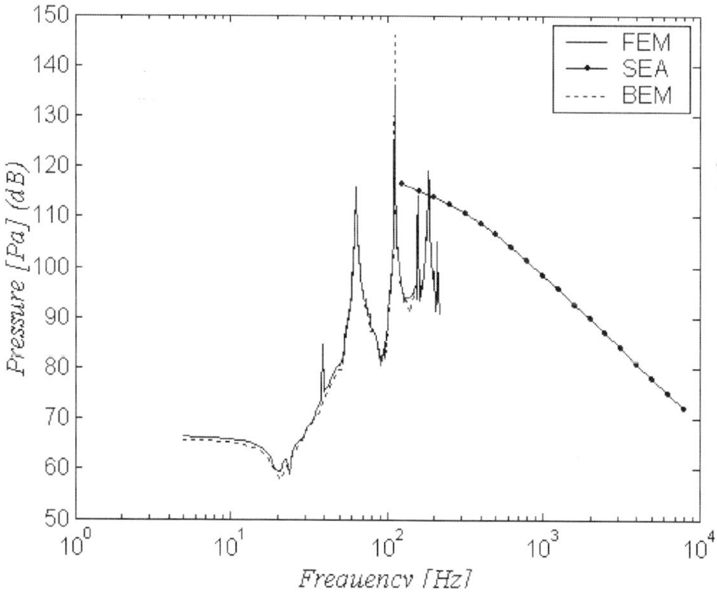

Figure 13. Acoustic Response Inside PLF

Considering the accuracy, advantages and drawbacks of the deterministic and statistical techniques, each of them is successfully applied in different frequency ranges. Such that, for the analyzed PLF, valid response results using deterministic techniques are assumed up to 150 Hz, while valid SEA results are assumed from 300 up 8,000 Hz. It is important to mention that in the "twilight zone" or medium frequency bandwidth (from 150 to 300 Hz), where deterministic models are inaccurate and present prohibitive computation time for the calculations and where the high modal density requirement is not yet accomplished for SEA, both results may be considered, as shows figure 13.

Model Validation

The fairing structure was positioned inside an acoustic chamber and excited with an ADF of 145 dB OSPL. Eight control microphones were positioned inside the reverberant chamber, which feedback the control system. Four measurement microphones were located in the acoustic cavity of the PLF. The measured space averaged SPL is compared with the theoretical acoustic responses, computed using the virtual prototypes (FEM and SEA models) (Fig. 14).

The calculated internal acoustic frequency response function shown in figure 13 may be transformed into 1/3 octave band responses to be compared with the experimental (measured) results. Figure 14presents the 1/3 octave comparisons for the frequency bandwidths ranging from 31.5 up to 8,000 Hz. It can be noticed, that experimental and calculated low-frequency responses have good agreement, presenting more significant differences only on the 1/3 octave bands 31.5, 40 and 50 Hz. This is because the low-frequency modes of the acoustic chamber are not well excited. However, in the regions where the cavity response is dominant (63 Hz and 112 Hz), differences are pretty small.

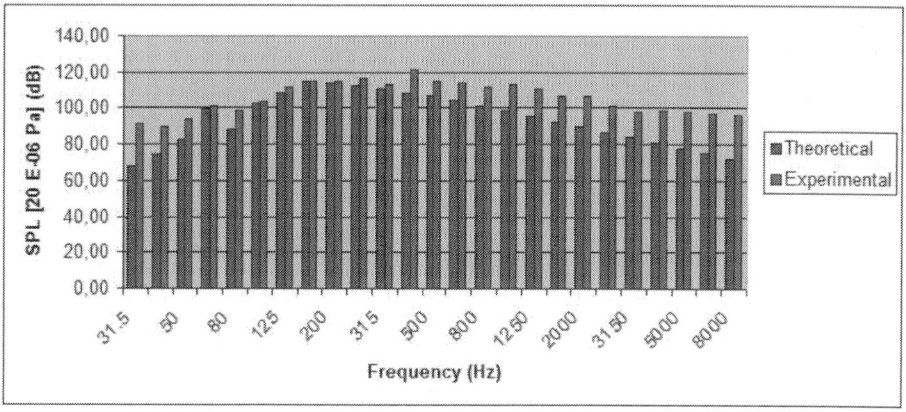

Figure 14. High-frequency theoretical X experimental comparison

For the higher frequencies, a more reasonable comparison should be done using Power Spectral Density (PSD) [28], since SEA calculations may be interpreted as mean values of energetic response functions when averaged at a given frequency over an ensemble of similar systems, differently of peak values resulting from deterministic approaches. However, a qualitative comparison can be presented for 1/3 octave bands from 160 up to 8,000 Hz, since one keeps in mind that mean values and the predicted magnitudes yielded by SEA should under estimate the dynamic response with a certain (acceptable) variance. Valid SEA results are assumed from 300 up 8,000 Hz, since the minimum five modes by bandwidth (modal density) requirement becomes true starting from 300 Hz.

At the beginning phases of space projects, the assessment of the effect of using different passive techniques for acoustic environment alleviation to be applied to PLF is an important issue. One of the main applications of numerical control prediction is the decision, still in the early product development phase, which design version is the most appropriate from the noise control point of view. By introducing the concept of sensitivity analysis, product development can be performed in a more systematic way. In order to predict the efficiency of a NCT, one compares the effects of design modifications.

In this framework, different blanket layers are implemented on the PLF elasto-acoustic virtual prototype and the effects of these NCT implementations are assessed. Since blankets acoustic absorption depend on certain material parameters, two blanket modeling approaches are assessed as follows: material physical Biot´s parameters as density, porosity, resistivity, tortuosity, viscous and thermal characteristic lengths, given by the blanket manufacturer and measured normal incidence absorption coefficients of material samples.

Furthermore, the influence of the NCT thicknesses and the presence of air-gaps between blankets are analyzed. Biot´s parameters and absorption coefficient approaches are implemented in the coupled elasto-acoustic SEA model of the PLF.

For the Biot´s parameters approach, an explicit model of the inserted material is considered, based on the physical properties of individual layers, which are accounted in the SEA model. Six types of glass wools are analyzed and the SPL inside the fairing are calculated. The wools' densities are given in pounds per cubic feet (pcf - 1 lb/ft^3 = 16.02 Kg/m^3). A thickness of 7,62 cm is adopted for almost all glass wools, but the two 1.2pcf glass wools, that presents particular behavior, which adopted thicknesses were 0.19 cm and 0.38 cm. The best performing material is chosen and a comparison between different thicknesses and percentages of layered surfaces of the fairing is done, considering the final weight of the applied NCT. The materials used were glass wools described in Table 5. The wools' Biot's parameters can be found in [4].

Table 5. Glass wools used

Material characteristcs						
Density (pcf)	0.34	0.42	0.6	1.2	1.2	1.5
Thickness (inches)	3	3	3	0.75	1.5	3

On the other side, the measured absorption coefficient of multi- and single-layered samples of glass wools of 0.42 and 1.0 pcf were considered. According to [26], air gaps between materials increase the acoustic absorption at low-frequencies. For this case, samples with two different air gaps are positioned into a Kundt tube. The single-layered samples are 3.50 cm thick, while combinations are done with samples of 1.75 cm thick. Other configurations were assembled with air gaps of 1.0 and 3.0 cm between samples. Figure 15 shows the sample combinations. All the measured absorption coefficients are shown in Fig. 16. These absorption coefficients are assigned on the fairing vibro-acoustic model and SPL are calculated.

The PLF acoustic responses for different NCT configurations are shown in Fig. 17. Notice that the insertion of the 0.34pcf glass wool – 7.62 cm yields almost 20 dB of attenuation (chosen as the best performing material). The assessment of the thickness influence is done by assigning 0.34pcf glass wools of 7.62, 10.16 and 12.7 cm thicknesses, with total NCT weights of 3.90, 5.30 and 6.60 Kg, respectively. Figure 18 shows the internal SPL one-third octave distribution, as well as the OSPL.

Figure 19 shows SPL and OSPL from 50 to 8,000 Hz, for the NCT described in Fig. 16, without air gaps. A 3.50 cm double-layered blanket (0.42pcf/1.0pcf) is compared with two single-layered NCT. Notice in this figure that NCT decrease the internal OSPL from 132 dB to 128 dB. Figure 20 shows that single-layered treatment with 1.0 pcf and air gaps presented better results. One can see the air gap effect, since the SPL close to 100, 315 and 500 Hz are higher mainly when the 1.0 pcf material with 3.0 cm air gap is applied. The calculations yielded 127.5 dB OSPL inside the fairing cavity. This means that a gain of approximately 3.0 dB at 100 Hz bandwidth can be obtained, yielding an overall gain of 1.0 dB, approximately. However, air gap installation can be limited, due to fairing internal space. In this case, it is preferable to install the blanketed treatment distant from the panels by small air gaps, instead of bonded, once this installation configuration presents higher transmission loss [10].

Figure 15. Double-layered 0.42pcf (yellow)/1.0pcf (orange), 1.75 cm thick each.

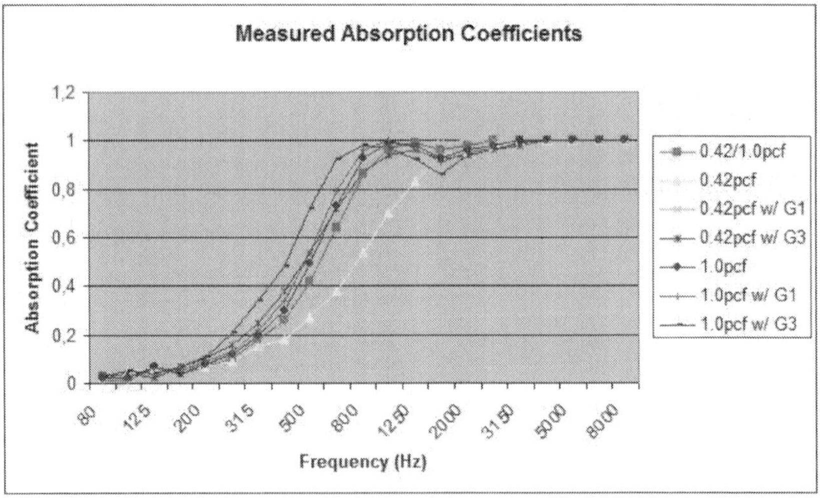

Figure 16. Measured absorption coefficients*G1 and G3: air gaps of 1.0 cm and 3.0 cm

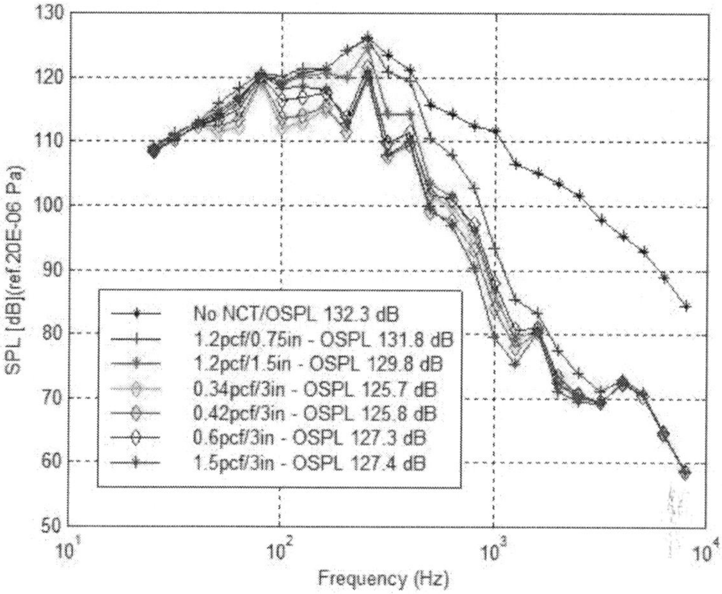

Figure 17. SPL for different blankets

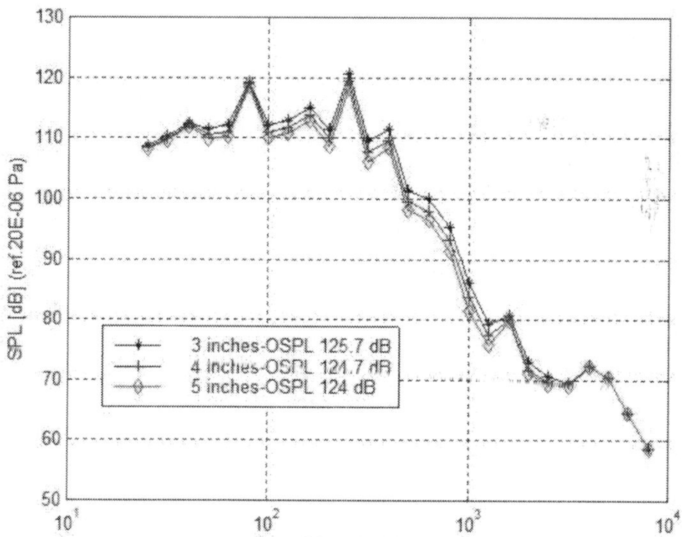

Figure 18. SPL for different thicknesses

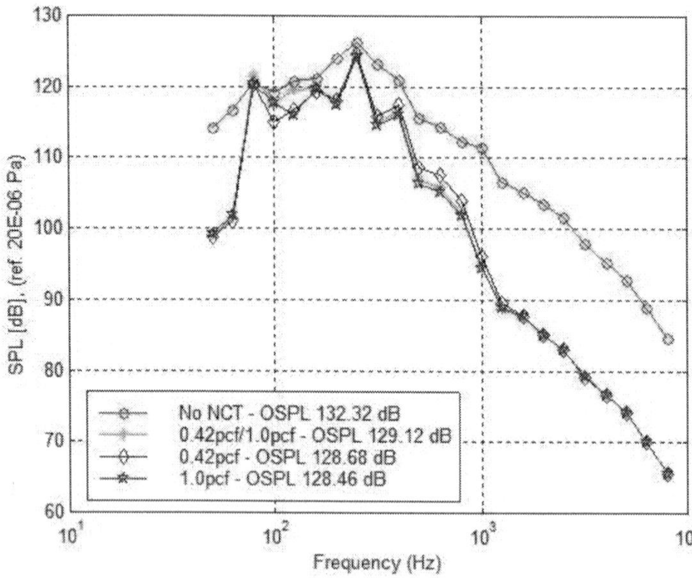

Figure 19. SPL for NCT without air gap.

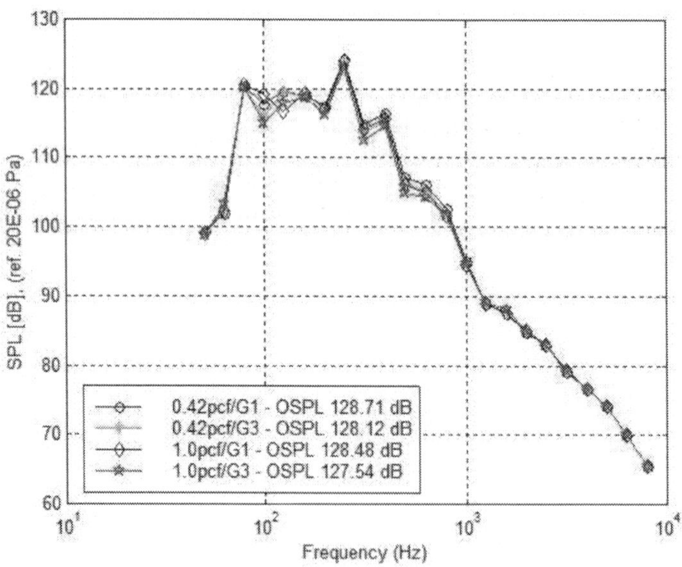

Figure 20. SPL for NCT with air gaps (*G1 and G3: air gaps of 1.0 cm and 3.0 cm)

COMBUSTION INSTABILITIES OF LIQUID PROPELLED ROCKET ENGINES DUE TO CHAMBER COMBUSTION ACOUSTICS

Combustion instabilities have been present in the development of LPRE over the last decades. There are basically three types of combustion instabilities: low-frequency (LF), medium-frequency (MF) and high-frequency (HF). LF instabilities, also called chugging, are caused by pressure interactions between the propellant feed system and the combustion chamber. MF instabilities, also called buzzing, are due to coupling between the combustion process and the propellant feed system flow. The HF instabilities are the most potentially dangerous and not well-understood ones. It occurs due to coupling of the combustion process and the chamber acoustics [27].

The presence of acoustic combustion instabilities must be considered still in development phase, although combustion instabilities can be clearly identified only during firing tests. In [13], it was described that instability can be verified when the power spectrum of the acoustic pressure levels, measured during burning tests, is analyzed. When an oscillation is observed, i.e., combustion instability, sound pressure peaks with well-defined magnitudes summed to the background noise are present. These peaks are correlated to the resonance frequencies of the combustion chamber. This phenomenon can cause instabilities and consequent unexpected behavior such as efficiency loss or even explosion of the engine. In this framework, the engine acoustic cavity characterization becomes an important issue to be investigated.

Acoustic behaviour of chambers is usually determined by doing cold tests measurements (without combustion). Acoustic dynamics in combustion environments are obtained by shifting the cold test resonant frequencies by a scalar factor defined by the ratio of sound velocity at the cold test temperature and at real operation temperature [15].

In view of attenuating acoustic pressure oscillations inside combustion chambers, reactive techniques as Helmholtz Resonators (HR), among others, are widely used ([13] and [16]). These devices are specially designed to attenuate oscillations at discrete resonance frequencies (pure tones). HR have been applied as combustion stabilization devices for solid motors and liquid rocket engines, with success. It could be noted in literature that they are used in a set of dozens or even hundreds in each chamber cavity, distributed along the walls or in a single row along the injector periphery [28].

This item describes a procedure for cold test acoustic characterization of LPRE combustion chambers. Firstly, the acoustical dynamic characterization of a combustion chamber is done and a typical longitudinal resonant frequency is chosen to be attenuated. A HR is designed (tuned at the chosen frequency) and applied to the mock up face plate. A LPRE mock-up [14] was used as experimental model. This test rig faithfully represents the internal acoustic cavity of the original engine. This procedure is followed by doing virtual prototypes of the combustion chamber. The acoustic natural frequencies and mode shapes are numerically calculated by a FEM model and validated through acoustic experimental modal analysis [29].

Experimental Acoustic Modal Analysis (EAMA)

Experimental Modal Analysis (EMA) is a well-applied technique in structure dynamics. However, due to the development of commercial acoustic sources, EAMA can be a suitable choice in view of extracting the acoustical Frequency Response Functions (FRF). In addition, the mathematical approach of structures modal parameters extraction can be applied to acoustic systems. [29].

In order to check the mutual orthogonality among modes from a modal model and to compare modes between different modal models (i.e., experimental and numerical solutions), the Modal Assurance Criterion (MAC) was used. This criterion indicates the degree of linear dependence between two eigenvectors and can be described as Eq. (7) [30].

$$MAC_{ijm} = 100 \cdot \frac{\left| \{\phi_{im}\}^T \{\phi_{jm}^*\} \right|^2}{\{\phi_{irm}\}^T \{\phi_{im}^*\} \{\phi_{jm}\}^T \{\phi_{jm}^*\}} \tag{7}$$

where: indexes *i* and *j* denotes modes obtained by different methods.

Helmholtz Resonator

Helmholtz Resonators are widely applied in order to suppress or attenuate the acoustic pressure inside cavities, rooms and other volumes. A HR consists of a small volume connected to a bigger cavity (the combustion chamber, in this case) through an orifice by a flanged neck. The dimensions of the HR must be much smaller than the acoustic wavelength of interest, in order to consider the resonator as lumped elements coupled to a geometric discontinuity. The coupling condition is that the oscillatory volume flow in the neck is equal to that imposed on the fluid inside the cavity, neglecting the elastic property of the air in the neck [30].

A typical HR is shown in Fig.21 (left), being *d* the neck diameter, *D* the cavity diameter, *Vc* the volume cavity, *l* the neck length and *L* the cavity length. *P1* is the incident acoustic pressure and *P2* is the cavity pressure. The gas motion in the HR coupled in an acoustic cavity can behave equivalently to a mass-spring-dashpot system (Fig.21, centre). The system can be divided into three distinct elements. The fluid enclosed in the neck behaves as an uncompressible gas, and its mass correspond to the *m* element of the mechanical system. The air inside the cavity is compressible and stores potential energy, representing the mechanical stiffness *k*. The mechanical damping element (*c*) is represented by two factors: (i) the open-end of the neck radiates sound, introducing a radiation resistance and (ii) the gas movement in the neck introduces a viscous resistance. Considering the electrical analogue (Fig.21right), the acoustic compliance *C* (analogous to electrical capacitance) is related to the stiffness of the air in the cavity, the acoustic inertance *M* (analogous to electrical inductance) is associated to the inertia element (mass) and the acoustic resistance *R* (analogous to electrical resistance) is related to the dissipative components stated above.

Considering that the gas beyond the end of the neck moves as a whole with the gas inside the neck, it is necessary to use an effective length *leff* which is bigger than the true length *l* of the neck [19]. The effective length *leff* is obtained by adding a mass end correction δ, which is empirically determined. In [28], it was presented a complete set of recommended equations for mass end correction, depending on the adopted considerations. For the purpose of this work, the appropriated equation is defined by:

$$\delta = 0.85d\left(1 - 0.7\sqrt{AR}\right) \text{for } AR < 0.16 \tag{8}$$

where *AR* is the Area Ratio (A_n / A_c), being A_n and A_c the neck cross-sectional area and the cavity cross-sectional area, respectively. The effective length is calculated as $l_{eff}=l+\delta$.

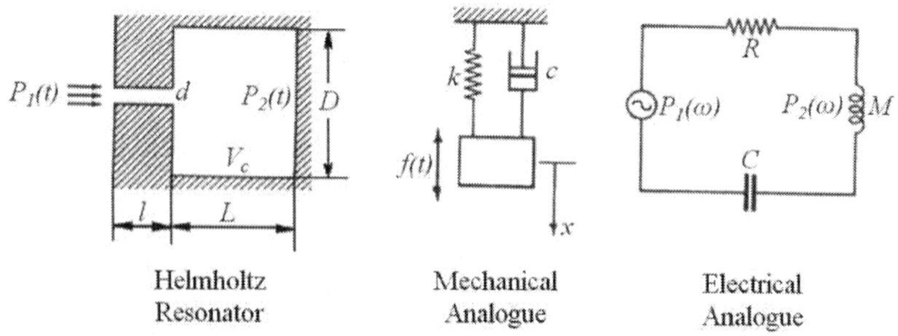

Figure 21. Helmholtz Resonators scheme and its analogues

The definition of acoustic inertance (*M*) applied to the Helmholtz resonator gives:

$$M = \frac{m}{A_n^2} = \frac{\rho l_{eff} A_n}{A_n^2} = \frac{\rho l_{eff}}{A_n} \tag{9}$$

where ρ is the air density and *m* is the effective mass.

The acoustic compliance *C* is defined as the volume displacement that is produced by the application of unit pressure [19]. By applying this definition to HR, one obtains:

$$C = \frac{V_c}{\rho c^2} \tag{10}$$

where *c* is the velocity of sound.

The acoustic resistance in the neck (*R*) was approximated as the dissipation associated with viscous forces, considering the dynamic viscosity μ [28]:

$$R = \frac{8\pi\mu l}{A_n} \tag{11}$$

The acoustic impedance *Z* of the HR is:

$$Z = R + j\left(\omega M - \frac{1}{\omega C}\right) \tag{12}$$

As can be seen, the acoustic impedance is determined by the geometric and mechanical properties of the resonator. The resonance will occur when the acoustic reactance equals zero:

$$\omega M - \frac{1}{\omega C} = 0 \tag{13}$$

The resonance frequency can be determined by considering that the dimensions of the resonator are much smaller than the wavelength of interest:

$$\omega_0 = \sqrt{\frac{1}{MC}} = c\sqrt{\frac{A_n}{l_{eff}V_c}} \tag{14}$$

The resonance's sharpness of a HR can be quantified by its quallty factor *Q*, given by:

$$Q = \frac{\omega_0 M}{R} \tag{15}$$

Finite Element Model

The cavity of a LPRE combustion chamber was analyzed using FEM in configurations without and with resonators. The first was modeled using 11,136 linear solid hexahedral elements, 12,510 nodes (12,093 degrees of freedom) and the second was modeled using 38,052 linear solid tetrahedral elements, 7,493 nodes (7,399 degrees of freedom). Both meshes are shown in Fig. 22. The fluid is air at 15° C. The eigenfrequencies were calculated from 0 to 2,400 Hz. Nodal pressures on the openings were assigned to zero.

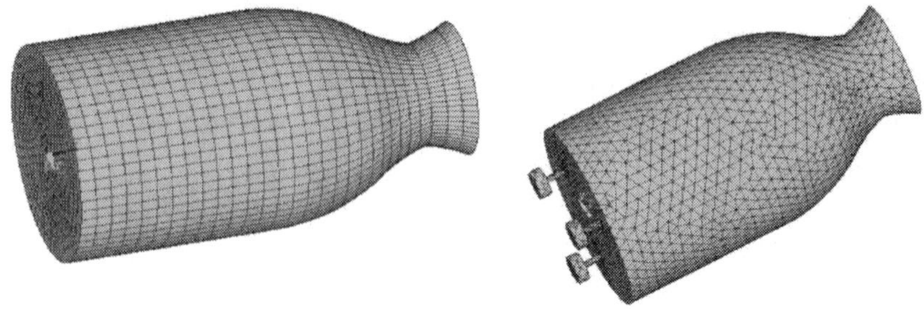

(a) (b)

Figure 22. a) Cavity without resonators; b) Cavity with resonators

Experimental Setup

Figure 23 shows the experimental setup. The mid-high frequency volume acceleration source is composed by a driver, a tube and a nozzle, where it is installed a volume acceleration sensor. This source produces a voltage signal proportional to the volume acceleration [m^3/s^2] variation, with a nominal frequency range of 200 up to 8,000 Hz. This source nozzle was installed in the mock-up plane surface as shown in Fig. 23.

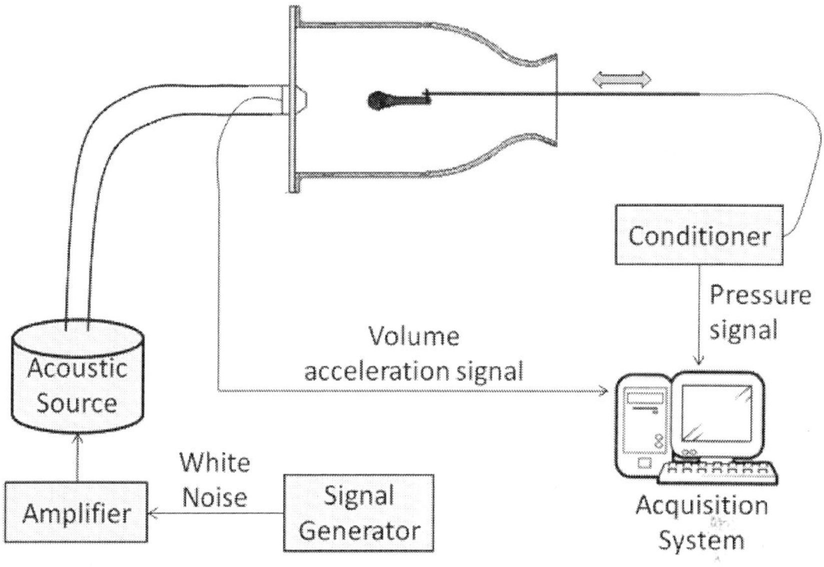

Figure 23. Measurement setup

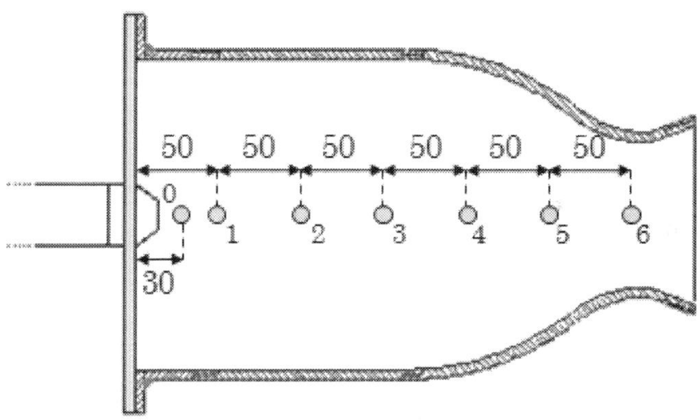

Figure 24. Source and microphone positions (distances in mm)

The chamber excitation was provided by a signal generator, a power amplifier and the source. The microphone was supported by a thin rod, placed in each measurement point inside the cavity. The pressure oscillations inside the cavity were captured by the microphone and registered by the data acquisition system. The volume acceleration source channel was settled as the reference channel. A white noise was used as excitation signal and the FRF were acquired at 7 points along the longitudinal axis (Fig. 24), being the point 0 the FRF driving point. The FRF were obtained by considering the volume acceleration as the excitation and the sound pressures as the responses. In order to make compatible theoretical x experimental comparisons, the volume velocity was assessed (instead of volume acceleration).

HR Design

The objective is to tune the HR resonance as the same frequency that must be attenuated. Due to construction facility; it was chosen a cylindrical shape to develop the HR. Not only the resonance frequency must be observed during the design process, but also several factors that influence directly the behavior of the HR:

- Resonance frequency of interest = 730 Hz (second longitudinal mode to be attenuated);
- Relation HR dimensions and wavelength λ (must be at least 10 times smaller than λ);
- Area Ratio (AR) must be smaller than 0.16, in order to assure an end mass correction
- Quality Factor (Q): monitored to be used in future designs (compare with other HR shapes);
- Constructive factors. Define such dimensions that can be feasible constructively.

The parameters used in the design were updated considering the room temperature (28 °C) observed during the experiment:: sound speed: $c = 348.3\ m/s$; air density: $\rho = 1.1839\ kg/m^3$; air dynamic viscosity: $\mu = 1.983x10^{-5}\ kg/ms$.

Considering the geometric dimensions in Eq. (14), one obtains the tuned resonance frequency as *726 Hz*, with a quality factor of $Q = 372.8$. The acoustic parameters were calculated $M = 1019\ kg/m^4$, $C = 4.72x10^{-11}\ s^2m^4/kg$, $R = 12468\ kg/sm^4$, using the Eq. (9), (10) and (11), respectively.

Three HR were manufactured in nylon. The presented measurement methodology was repeated in order to acquire the same FRF, considering the new configuration, with the resonators.

Results

The identified natural frequencies are summarized in Table 6. The transversal modes frequencies were identified only numerically. Considering the first four longitudinal modes, the maximum error comparing the numerical and experimental estimation of natural frequencies was 6.6%.

Figures 25a, 25c and 25e present the three first longitudinal numerical modes.

Table 6. Experimental and Numerical Natural Frequencies (Hz)

Mode	Experimental	Numerical	Error (%)
Longitudinal 1st	164.5	170.28	3.51
Longitudinal 2nd	730.85	721.9	1.23
Transversal 1st	-	1126	-
Transversal 2nd	-	1126	-
Longitudinal 3rd	1272.64	1296	1.84
Transversal 3rd	-	1382	-
Transversal 4th	-	1382	-
Longitudinal 4th	1689.52	1801	6.6

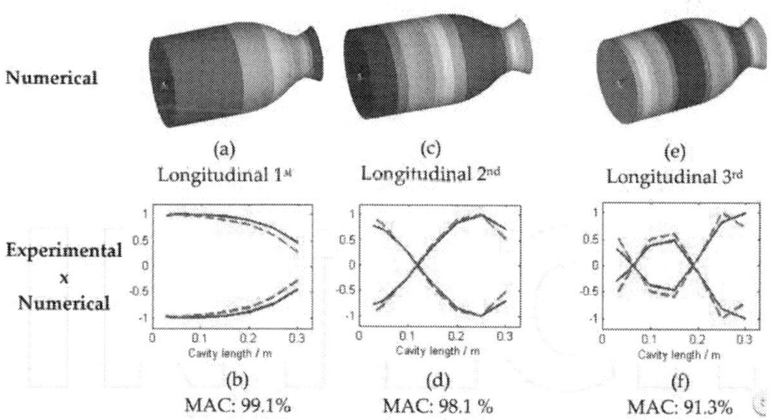

Figure 25. a) to f) Numerical modes and comparison between experimental (blue line) and numerical (red line) longitudinal mode shapes

Experimental versus numerical modes comparison, considering the normalized amplitude is also shown (Figs. 25b, 25d and 25f). In the numerical modes represented by the collor maps pictures (Figures 25a,25c and 25e) the nodal regions are in green. The MAC (Eq. (7)) is also presented.

Notice in Fig. 25 that MAC values are bigger than 91% for the three first modes. For the first and second modes MAC values reach about 99%. Figures 25d and 25f show that the nodes in these modes are almost at the same point. After the introduction of the HR, the attenuation of the second mode is clearly noted in Fig. 26a, when comparing measurement results of the original cavity. At least 9 dB of attenuation can be observed in the new configuration. The FRF with and without HR are almost the same, but the second mode region (about 730 Hz), where the HR is tuned.

Figure 26 depicts the behavior of the chamber with HR. Numerical mode shapes of the configuration with HR were plotted and correlated to each part of the experimental FRF. The Fig. 26b was zoomed from the squared highlight in Fig.26a (point 1). This allows visualize the entire system behavior.

In Fig. 26 it can be noticed that mode shapes *(c)* and *(d)* have similar behavior, but different natural frequencies: it can be realized that the pressures inside all resonators vary in phase. In this case, the pressure inside the chamber remains almost unchanged. In the mode shapes *(e)* and *(f)*, the resonators act close its tuned frequency. In these modes, the whole chamber behaves as a nodal region and the pressure inside the resonators varies out of phase.

As a result of the movement of the air mass inside the neck in the four modes represented in Figs. 26, the acoustic energy on the resonators behaves as expected, reducing the energy inside the chamber.

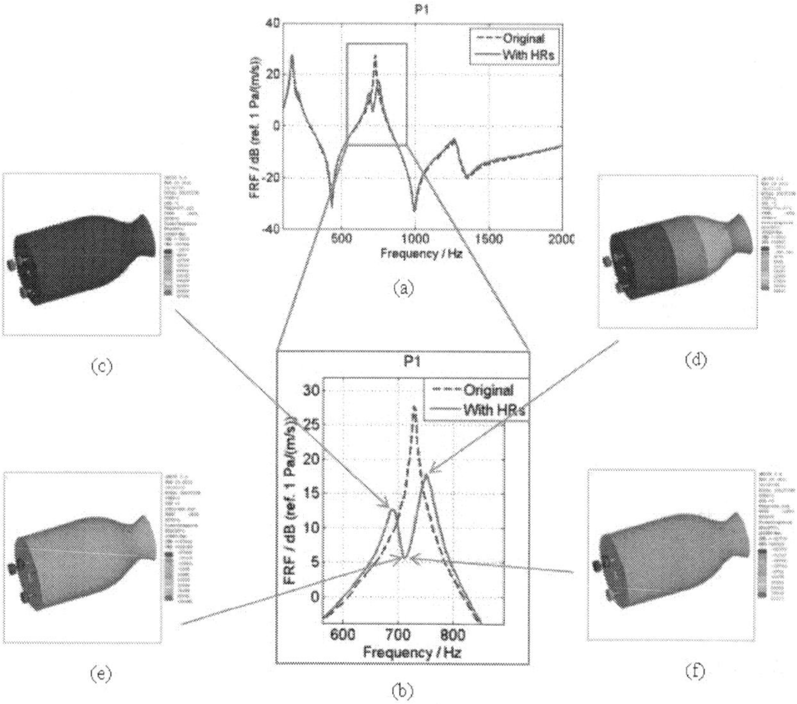

Figure 26. Numerical mode shapes near the resonances of the HR

CONCLUSIONS

This chapter presents three acoustic case studies applied on rocketry design.

Firstly, microphone protection devices design procedures for in-flight measurements are described. The modeling techniques using analytical and FEM numerical tools are presented as well as the validation acoustic testing procedures are presented. Good agreement among numerical and experimental results was obtained. A procedure to asses the SPL outer the launcher's structure by using the adapters' acoustic transfer functions and internal SPL measurements was also described.

Vibro-acoustic virtual prototypes were used to predict the acoustic response of a PLF cavity when excited by an ADF of 145 dB OSPL ranging from 5 to 8,000 Hz, generated at lift off.

Coupled deterministic techniques, using FEM/FEM and FEM/BEM, were applied to the fairing problem in a low-frequency band, considering accurate and efficient modeling techniques. The modal and semi-modal superposition techniques were applied to perform a FRA. In the higher frequencies, SEA coupling technique was applied to obtain the fairing acoustic responses in 1/3 octave bands.

The fairing was submitted to the lift off excitation in acoustic reverberant test and the internal acoustic pressure levels were measured. Experimental and numerical results show good agreement, except for the frequencies below 50 Hz and above 4,000 Hz.

The sensitivity analysis of acoustic blankets showed to be an effective tool for the development of the fairing NCT design. The effectiveness of a NCT considering its weight and performance can easily be evaluated using SEA, still in the development phase, when detailed subsystems are not required.

By analyzing many NCT configurations one can provide a library of performances and weights, important parameters that describe the ELV performance book. As in space industry the cost of a mission is a major issue, a trade-off between NCT weight and efficiency must be accounted. Acoustic testing in reverberant chamber may be conducted to validate the presented results and other porous-elastic materials may be investigated to complement the fairing NCT design library.

Finally, for the third case study, the use of a volumetric source in Experimental Acoustic Modal Analysis has important role in the process, once allows the accurate measurement of acoustic FRF.

The numerical model results were used as the basis for the HR design, in a first moment. In addition, numerical and experimental models were used to identify and localize, with a level of security, the node and maximum amplitude regions of each mode. The HR design seemed to be adequate, once it was verified an attenuation of 9 dB or bigger, depending of the location inside the chamber

REFERENCES

1. P. Jorge, Ravi. N. Arenas, Margasahayam, 2006 Noise and Vibration of Spacecraft Structures, Ingeniare. Revista Chilena de Ingeniería, 14 Nº 3, 251264
2. R. Pirk, P. Sas, W. Desmet, L. C. S. Góes, ". Vibro-Acoustic, of. Analysis, Vehicle. the, Launcher´s. . V. L. S. Sattelite, Ph. Fairing", D. Thesis, 2003 Technological Institute of Aeronautics (ITA), São José dos Campos, Brazil.
3. R. Pirk, L. C. S. Góes, 2006 "Acoustic Theoretical x Experimental Comparison of the Brazilian Satellite Launcher Vehicle (VLS) Fairing", Proceedings of the ISMA 2006, Leuven, Belgium.
4. R. Pirk, Carlos. Souto, A. d´, 2009 Implementation of Acoustic Blankets to the VLS Fairing- A Sensitivity Analysis Using SEA, Proceedings of the ISMA 2008, Leuven, Belgium.
5. J. M. Scott, B. G. Jay, C. B. Robert, 1996 General Environmental Specification for STS and ELV- Payloads, Subsystems and Components, GEVS-SE, Rev. June 1996, NASA Goddard Space Flight Center, Maryland, 20771.
6. Souto and Pirk, 2009 Estimation of the external sound pressure levels generated during VLS- 1 lift off, Institute of Aeronautics and Space, Technical Report RT 001/AIE/R/2009).
7. Pirk et al. 2010 Alternative External Acoustic Loads Estimation of the Brazilian Satellite Launcher (VLS-1), Using In-Flight Experimental Data, International Congress on Sound and Vibration, Rio de Janeiro, Brazil
8. R. M. Glaese, E. H. Anderson, 2003 "Initial Structural-Acoustic Modeling and Control Results for a Full-Scale Composite Payload Fairing for Acoustic Launch Load Alleviation". http://www.csaengineering.com/techpapers/technicalpaperspdfs/CSA1999_ Initial Structural
9. K. Weissman, M. E. Mc Nelis, W. D. Pordan, 1994 "Implementation of Acoustic Blankets in Energy Analysis Methods with Application to the Atlas Payload Fairing, Journal of the Institute of Environmental Sciences- IES, July/August 1994.
10. I. S. Bolton, 2005 "Porous Materials for Sound Absorption and Transmission Control", Proceedings of the 2005 Congress and Exposition on Noise Control Enginnering, INTERNOISE 2005, Rio de Janeiro, Brazil.
11. H. Defosse, M. A. Hamdi, ". Vibro-acoustic, of. study, V. Ariane, during. launcher, Proc. lift-off", of Ariane V launcher during lift-off", Proc. of Inter-Noise 2000 1 9-14 (Nice, 2000).
12. F. E. C. Culick, Combustion Instabilities in Liquid Rocket Engines: Fundamentals and Control. California Institute of Technology, 2002
13. Burnley, V.S.; Culick, F.E.C., The Influence of Combustion Noise on Acoustic Instabilities, Air Force Research Laboratory, OMB No 0704-0188, 1997.
14. R. Pirk, C. d´, A. Souto, D. D. Silveira, C. M. Souza, 2010 Liquid rocket combustion chamber acoustic characterization, Journal of Aerospace Technology and Management.

15. E. Laudien, Others, 1994 Experimental Procedures Aiding the Design of Acoustic Cavities, DASA- Deutsche Aerospace AG, Liquid Rocket Engine Combustion Instability, Progress in Astronautics and Aeronautics, Chapter 14, 169
16. M. S. Natanzon, 1999 Edited by Culick, F. E. C., California Institute of Technology
17. Troclet, B., Analysis of Vibro-acoustic Response of Launchers in the Low-Frequency and High-Frequency Range, Proc. of NOVEM 2000 (Lyon, France, 2000).
18. V. I. Khlybov, S. A. Mak′hankov, 2009 The Preparation to Experimental Studies and Acoustic Sensors Testing Program Development, Technical Report.
19. Lawrence. E. Kinsler, Austin. R. Frey, of. Fundamentals, John. Acoustics, Wiley, 2nd. Sons, edition, 1962EO
20. W. Desmet, D. Vandepitte, 2001 Finite Element Method in Acoustics, Course Graduate School in Mechanics- Advanced Acoustics, Katholieke Universiteit Leuven
21. J. Coyette-P, C. Lecomte, K. Meerbergen, 1997 Treatment of Random Excitations using SYSNOISE Rev. 5.3.1 - Documentation Theoretical Manual, Users Manual and Validation Manual, LMS Numerical Technologies NV.
22. J. Klos, et, 2003 Sound Transmission Through a Curved Honeycomb Composite Panel, AIAA Conference Paper AIAA-20033
23. Desmet, W, Vandepite, D., Boundary Elements in Acoustics, Katholieke Universiteit Leuven (KUL), 2001.
24. Lyon and DeJong, Theory and Application of Statistical Energy Analysis, Second Edition, 1995
25. T. Thinh, Modeling. Vibro-acoustic, of. Study, Delta. I. I. 1. the, 10 10 Composite Fairing, Journal of the IEST, November/December 1999.
26. J. F. Allard, 1993 Propagation of Sound in Porous Media- Modeling Sound Absorbing Materials, Elsevier Applied Science, London and New York.
27. G. P. Sutton, O. Biblarz, 2001 "Rocket Propulsion Elements", New York, John Wiley & Sons.
28. Santana Jr, A. , Investigation of Passive Control Devices to Suppress Acoustic Instability in Combustion Chambers. Thesis of doctor in science. Aeronautics Institute of Technology, São José dos Campos, Brazil, 2008
29. G. P. Guimarães, R. Pirk, C. A. Souto, L. C. S. Góes, Acoustic Modal Analysis of Cylindrical-Type Cavities. Proceedings of 8 8th Intern. Confer. on Structural Dynamics- EURODYN, Leuven, Belgium, 2011, 31603167 .
30. F. Fahy, Fundamentals of Engineering Acoustics. Academic Press. London, UK. 2001

CITATION

Rogério Pirk, Carlos d´Andrade Souto, Gustavo Paulinelli Guimarães and Luiz Carlos Sandoval Góes (2013). Acoustics and Vibro-Acoustics Applied in Space Industry, Modeling and Measurement Methods for Acoustic Waves and for Acoustic Microdevices, Prof. Marco G. Beghi (Ed.), ISBN: 978-953-51-1189-4, InTech, DOI: 10.5772/49966.

Index